Introduction to the Mathematics of Inversion in Remote Sensing and Indirect Measurements

S. TWOMEY

DOVER PUBLICATIONS, INC.
Mineola, New York

Copyright

Published in Canada by General Publishing Company, Ltd., 30 Lesmill Road, Don Mills, Toronto, Ontario.
Published in the United Kingdom by Constable and Company, Ltd., 3 The Lanchesters, 162–164 Fulham Palace Road, London W6 9ER.

Bibliographical Note

The present edition, first published by Dover Publications, Inc., in 1996, is an unabridged and slightly corrected republication of the work originally published in 1977 by Elsevier Scientific Publishing Company, Amsterdam, as No. 3 in their series "Developments in Geomathematics."

Library of Congress Cataloging-in-Publication Data

Twomey, S.
Introduction to the mathematics of inversion in remote sensing and indirect measurements / S. Twomey.
p. cm.
"An unabridged and slightly corrected republication of the work originally published in 1977 by Elsevier Scientific Publishing Company, Amsterdam as no. 3 in their series 'Developments in geomathematics' "—T.p. verso.
Includes bibliographical references and indexes.
ISBN 0-486-69451-8 (pbk.)
1. Remote sensing—Mathematics. 2. Matrix inversion. I. Title.
G70.4.T86 1996
621.36′78′0151—dc20 96-35745
CIP

Manufactured in the United States of America
Dover Publications, Inc., 31 East 2nd Street, Mineola, N.Y. 11501

Preface

Inversion problems have existed in various branches of physics and engineering for a long time, but in the past ten or fifteen years they have received far more attention than ever before. The reason, of course, was the arrival on the scene of large computers (which enabled hitherto abstract algebraic concepts such as the solution of linear systems of equations in many unknowns to be achieved arithmetically with real numbers) and the launching of earth-orbiting satellites which viewed the atmosphere from outside and which could only measure atmospheric parameters remotely and indirectly by measuring electromagnetic radiation of some sort and relating this to the atmospheric parameter being sought, very often implying a mathematical inversion problem of some sort.

It was soon found that the dogmas of algebraic theory did not necessarily carry over into the realm of numerical inversion and some of the first numerical inversions gave results so bad — for example, in one early instance negative absolute temperatures — that the prospects for the remote sensing of atmospheric temperature profiles, water vapor concentration profiles, ozone profiles, and so on, for a time looked very bleak, attractive as they might have appeared when first conceived.

The crux of the difficulty was that numerical inversions were producing results which were physically unacceptable but were mathematically acceptable (in the sense that *had* they existed they would have given measured values identical or almost identical with what was measured). There were in fact ambiguities — the computer was being "told" to find an $f(x)$ from a set of values for $g(y)$ at prescribed values of y, it was "told" what the mathematical process was which related $f(x)$ and $g(y)$, but it was not "told" that there were many sorts of $f(x)$ — highly oscillatory, negative in some places, or whatever — which, either because of the physical nature of $f(x)$ or because of the way in which direct measurements showed that $f(x)$ usually behaved, would be rejected as impossible or ridiculous by the recipient of the computer's "answer". And yet the computer was often blamed, even though it had done all that had been asked of it and produced an $f(x)$ which via the specified mathematical process led to values of $g(y)$ which were exceedingly close to the initial data for $g(y)$ supplied to the computer. Were it possible for computers to have ulcers or neuroses there is little doubt that most of those on which early numerical inversion attempts were made would have acquired both afflictions.

For a time it was thought that precision and accuracy in the computer were the core of the problem and more accurate numerical procedures were sought and applied without success. Soon it was realized that the problem was not inaccuracy in calculations, but a fundamental ambiguity in the presence of inevitable experimental inaccuracy — sets of measured quantities differing only minutely from each other could correspond to unknown functions differing very greatly from each other. Thus in the presence of measurement errors (or even in more extreme cases computer roundoff error) there were many, indeed an infinity of possible "solutions". Ambiguity could only be removed if some grounds could be provided, from outside the inversion problem, for selecting one of the possible solutions and rejecting the others. These grounds might be provided by the physical nature of the unknown function, by the likelihood that the unknown function be smooth or that it lie close to some known climatological average. It is important, however, to realize that most physical inversion problems are ambiguous — they do not possess a unique solution and the selection of a preferred unique solution from the infinity of possible solutions is an imposed additional condition. The reasonableness of the imposed condition will dictate whether or not the unique solution is also reasonable. There have been many advances in mathematical procedures for solving inversion problems but these do not remove the fundamental ambiguity (although in some instances they very effectively hide it). Our measurements reduce the number of possible solutions from infinity to a lesser but still infinite selection of possibilities. We rely on some other knowledge (or hope) to select from those a most likely or most acceptable candidate.

S. TWOMEY

Contents

List of frequently used symbols and their meanings

$\mathbf{A}$ usually a matrix, often the matrix of quadrature coefficients representing the numerical equivalent of the integral transform $\int K(y, x) f(x)\,dx$

$\mathbf{B}$ a matrix

B Planck radiance in radiative transfer

$\mathbf{C}$ usually the covariance matrix with general element $\int K_i(x) K_j(x)\,dx$

$\mathbf{D}$ a diagonal matrix other than the identity

$\mathbf{E}$ the matrix with general element $\langle \epsilon_i \epsilon_j \rangle$

$\mathbf{I}$ the identity matrix; also used as a scalar quantity for intensity in radiative transfer

$J_m(x)$ the Bessel function of the first kind

$K(y, x)$ the kernel function of an integral transform

$K_i(x)$ the kernel function when y takes discrete values y_i $(i = 1, 2, \ldots)$

$\mathbf{L}$ usually the matrix or operator associated with a linear transformation

$P_m(x)$ Legendre polynomial

Q usually a quadratic form

$\mathbf{R}$ usually a rotation matrix

$\mathbf{U}$ a matrix containing the eigenvectors $\boldsymbol{u}_1, \boldsymbol{u}_2$, etc., in its columns

$\boldsymbol{e}$ a vector with every element unity

$\boldsymbol{e}_k$ a vector with kth element unity, all others being zero

$f(x)$ a function of x, usually that which is being sought in inversion contexts

$\boldsymbol{f}$ $f(x)$ in tabular (vector) form

$g(y)$ the given or measured function in an inversion context

$\boldsymbol{g}$ $g(y)$ in tabular (vector) form

$\boldsymbol{k}(x)$ a vector with elements $K_1(x)$, $K_2(x)$, etc.

t, $t(x)$ transmission (transmittance) in radiation contexts

$\boldsymbol{u}$ an eigenvector

γ Lagrangian undetermined multiplier

$\delta(x)$ the Dirac delta function

ϵ error term

$\boldsymbol{\epsilon}$ vector of error terms

θ direction of propagation measured from the zenith

κ absorption coefficient

λ eigenvalue of a matrix; wavelength (in radiation contexts)

μ $\cos\theta$

$\boldsymbol{\xi}$ an array of arbitrary coefficients

τ optical thickness

ϕ azimuth angle

$\hat{\phi}(x)$ a normalized function of x, usually one of an orthonormal set

$\boldsymbol{\phi}(x)$ a set of functions $\phi_1(x)$, $\phi_2(x)$, etc., collected into a vector

ω angular frequency; dependent variable after Fourier transformation

$\bar{\omega}_0$ single scattering albedo

$\langle \ldots \rangle$	denotes averaged quantities (an overhead bar is also used for averaging)
$\hat{}$	cap symbol denotes normalization
$\lvert \ldots \rvert$	represents the norm or magnitude of the quantity (scalar, function, vector, etc.) contained between the bars
$\lVert \ldots \rVert$	representation of a matrix when the compositions of the elements is to be explicitly indicated (see Chapter 3, section 3.2)
$\ldots^*$	The asterisk is used throughout to represent transposition. When $\boldsymbol{v}$ is a column vector its transpose $\boldsymbol{v}^*$ is a row vector
$\cdot$	The dot symbol is used occasionally for the inner or dot product of two functions or vectors. For vectors the notation $\boldsymbol{v}^*\boldsymbol{u}$ is equivalent to $\boldsymbol{v} \cdot \boldsymbol{u}$ and is preferred

CHAPTER 1

Introduction

Many, indeed most, of our everyday methods of measurement are indirect. Objects are weighed by observing how much they stretch a spring, temperatures measured by observing how far a mercury column moves along a glass capillary. The concern of this book will not be with that sort of indirectness, but with methods wherein the distribution of some quantity is to be estimated from a set of measurements, each of which is influenced to some extent by all values of the unknown distribution: the set of numbers

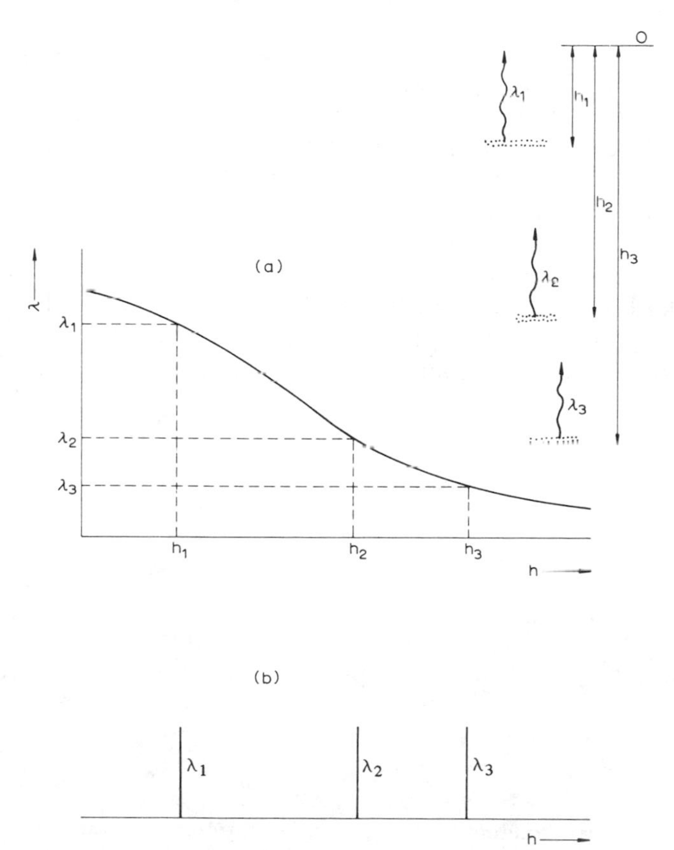

Fig. 1.1. Schematic diagram for an indirect sensing measurement such as satellite-based atmospheric temperature profile measurement.

which comprises the "answer" must be unravelled, at it were, from a tangled set of combinations of these numbers.

An illustration may help to make clear what is meant. Suppose that any level in the atmosphere radiated infrared energy in an amount which depended in a known way on its temperature (which indeed is true), and suppose also that each level radiated energy only at a single wavelength characteristic of that level and emitted by no other level (which is not at all true). Clearly, we could place a suitable instrument in a satellite above the atmosphere and obtain the temperature at any level by measuring the infrared intensity at an appropriate wavelength, which could be obtained from a graph of wavelength against height as in Fig. 1.1a. From Fig. 1.1a we could construct Fig. 1.1b which shows for an arbitrary set of wavelengths $\lambda_1, \lambda_2, ..., \lambda_n$ the relative contribution of various height levels to the measured radiation at that wavelength. Because of the imagined unique relation between height and wavelength, these contributions are zero everywhere except at the height which Fig. 1.1a indicates to be radiating at that particular wavelength.

If the ideal relationship existed the graph of Fig. 1.1b would only be an awkward and less informative version of the graph of Fig. 1.1a. But suppose now we wish to describe the situation where some blurring exists — where the level given in the figures contributes most strongly to the corresponding wavelength, but the neighboring levels also contribute to an extent that diminishes as one moves farther from the level of maximum contribution. This cannot readily be shown in a figure similar to Fig. 1.1a, but is easily shown by the simple modification to Fig. 1.1b which is seen in Fig. 1.2. Each relative contribution is still in a sense localized around a single level, but it is no longer a spike but rather a curve which falls away on both sides from a central maximum. The wider that curve the more severe the blurring or departure from the one-to-one correspondence between wavelength and height.

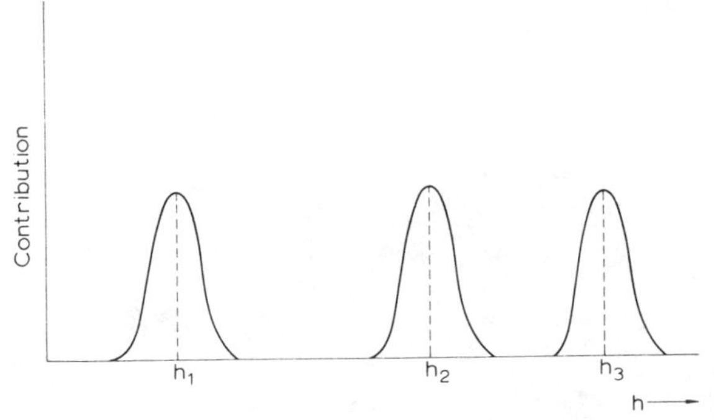

Fig. 1.2. A version of Fig. 1b, corresponding to the practical situation where the measured radiation does not originate from a single level, but where different levels contribute differently.

1.1. MATHEMATICAL DESCRIPTION OF THE RESPONSE OF A REAL PHYSICAL REMOTE SENSING SYSTEM

To describe the blurring process just mentioned in a mathematical way is not difficult. Let $f(x)$ be the distribution being sought and let $K(x)$ represent the relative contribution curve — since there is one such curve for every wavelength we must add a subscript making it $K_i(x)$ or, alternately, we may specify the wavelength dependence by considering the relative contributions to be a function of the two variables and writing it as $K(\lambda, x)$. In most situations the subscript notation accords better with the facts of the practical situation where a *finite* number n of wavelengths $\lambda_1, \lambda_2, ..., \lambda_n$ are utilized. The difference is only one of notation; we have $K(\lambda_i, x) \equiv K_i(x)$. A measurement at wavelength λ_i involves radiation not just from a height (say x_i) at which $K_i(x)$ is a maximum, but also from the neighboring region within which the contribution and $K_i(x)$ do not vanish. If the interval between x and $x + \Delta x$ contributes to the ith measurement the amount $f(x)\ K_i(x)\ \Delta x$ then the total measured radiation is clearly $\int K_i(x)\ f(x)\ dx$; the limits to be placed on this depend on the exact circumstances. In almost all experiments x cannot be negative and in many experiments it does not exceed an upper limit X. We may find indirect sensing problems wherein the relevant integral is either $\int_0^\infty K_i(x)\ f(x)\ dx$ or $\int_0^X K_i(x)\ f(x)\ dx$. In the latter case it is always possible to redefine the independent variable from x to x/X and thereby to make the integral $\int_0^1 K_i(x)\ f(x)\ dx$.

Convolution

If the blurring functions or contribution functions shown in Fig. 1.2 are always identical in shape, differing only in position along the x-axis, then, if we define $K_0(x)$ to be the function centered at the origin ($x = 0$), we can write $\int K_i(x)\ f(x)\ dx$ in the somewhat different form $\int K_0(x - x_i)\ f(x)\ dx$, x_i being the value of x at which $K_i(x)$ has a maximum. Since $K_0(x)$ attains a maximum (by definition) at $x = 0$, $K(x - x_i)$ attains a maximum at $x = x_i$ and so is nothing more than $K_0(x)$ displaced along the x-axis by an amount x_i (see Fig. 1.2). The quantity $\int K_0(x - x_i)\ f(x)\ dx$ is clearly a function of x_i, it depends also on the shape of $K_0(x)$ and $f(x)$ and is known as the "convolution" or "convolution product" of the two functions. Sometimes the convolution of $K_0(x)$ and $f(x)$ is represented by a product notation, such as $K_0 {}^* f$. Convolutions play an important role in Fourier and Laplace transform theory. They also have a practical significance: when, for example, an optical spectrum is scanned by a spectrometer, the result is not the entering spectral distribution: every spectrometer has a slit or some other way of extracting energy of a prescribed wavelength, but there is no practical way to achieve the mathematical abstraction of a slit which is infinitesimally narrow, so the very best a spectrometer can do is to pass a desired wavelength λ plus some neighboring wavelengths in a narrow band centered at λ. The curve describing how the relative contribution varies as we go away from the desired "central wavelength" λ is known as the "slit function" of the instrument. If the slit function is $s(u)$ where u denotes the difference in wavelength away from the central wavelength λ, then an initial spectral distribution $E(\lambda)$ will contribute to the spectrometer measurement at wavelength λ not $E(\lambda)$ but $\int E(\lambda')\ s(\lambda' - \lambda)\ d\lambda'$, a convolution of E and s which results in a somewhat smoothed, blurred version of $E(\lambda)$, the degree of blurring depending on the width of slit function s. In situations where attempts are made to "unblur" or sharpen the measured spectra this is often called "deconvolution". It is evident that deconvolution is very closely related to the inversion of indirect sensing measurements, being merely a special case where the shape of the functions $K_i(x)$ are unchanged apart from displacement along the x-axis.

We may note in passing that processes such as the formation of running means from a sequence of numbers are also convolutions. If, for example, we take the following sequence of values for a variable which we shall call y:

$$y_1 = 0, \quad y_2 = 1, \quad y_3 = 3, \quad y_4 = 6, \quad y_5 = 10, \quad y_6 = 10,$$
$$y_7 = 9, \quad y_8 = 7, \quad y_9 = 7, \quad y_{10} = 10$$

If now we form running means $\bar{y}$ by averaging the y's three at a time, so that, for example, $\bar{y}_5 = \frac{1}{3}(y_4 + y_5 + y_6)$, then the result (given to two decimal places): $\bar{y}_1$ = undefined, $\bar{y}_2 = 1.33$, $\bar{y}_3 = 3.33$, $\bar{y}_4 = 6.33$, $\bar{y}_5 = 8.67$, $\bar{y}_6 = 9.67$. $\bar{y}_7 = 8.67$, $\bar{y}_8 = 7.67$, $\bar{y}_9 = 8$, $\bar{y}_{10}$ = undefined, represents a smoothing of the y data and can be written as the convolution of the y data, regarded as a suitably defined distribution of y with respect to a continuous variable x, with a smoothing function which is rectangular in shape, of unit area and width 3 (i.e. amplitude 1/3).

Delta function

To convert the tabulated y-values in the last paragraph to running means is a very simple process, but to describe it rigorously by an integral we have a problem, since the values y_1, y_2, y_3, etc., were envisaged to be attained exactly at $x = 1$, $x = 2$, $x = 3$, etc.; if we define the distribution $y(x)$ to be zero everywhere except at $x = 1, 2, 3$, etc., then the integral $\int y(x)\, dx$ is evidently zero, whatever the interval and we can easily show that any bounded function w of x will give a vanishing integrated product $\int y(x)\, w(x)\, dx$. To avoid this problem the so called "delta function" is useful. This is not a *function* in the strictest mathematical sense. It may be regarded as a distribution which is zero everywhere except in an infinitesimal interval within which it becomes infinite in such a way that its integral $\int_{-\infty}^{\infty} \delta(x)\, dx$ remains finite and in fact unity. It may be regarded as the limit of a sequence of discontinuous functions (e.g. rectangular distributions) or even of a sequence of continuous functions $[(\sin mx)/mx]$ for $m \to \infty$; the delta function corresponds to a slit of infinitesimal width.

The delta function is useful in a number of contexts and it may be noted that convolution with a delta function represents an identity operation — the function convoluted with a delta function is unchanged; we merely "reread" the graph of $f(x)$ with infinite precision and do not blur it in any way.

Returning now to Fig. 1.1 we may say that the measurements in the ideal situation portrayed in Fig. 1.1 give us $\int K_i(x)\, f(x)\, dx$ with $K_i(x)$ the delta function $\delta(x - x_i)$, so that we "pick off" the value of $f(x_i)$ without any loss in precision (assuming the measurements are themselves precise). In the situation portrayed in Fig. 1.2 $K_i(x)$ has a relatively narrow profile, and the measurements give us a blurred convolution of $f(x)$ which still resembles $f(x)$ but may be sharpened by a suitable deconvolution process — in other words, we have directly and without inversion an approximation to $f(x)$ but it may be possible to improve it by some inversion method.

But the situation may be worse: the $K_i(x)$ may remain non-zero over the full range of x, possessing single maxima at x_1, x_2, x_3, etc., but not falling off quickly away from maximum (Fig. 1.3) — or they may even possess a common maximum (often at $x = 0$, as shown in Fig. 1.4), or they may have multiple local maxima. In such situations the measurements may not give directly any idea of what $f(x)$ looks like and it is evident that the problem of inversion (i.e. obtaining $f(x)$ or a useful approximation or estimate for $f(x)$) from measurements of $\int K_i(x)\, f(x)\, dx$ becomes more formidable.

In mathematical terms the inversion problem is that of finding $f(x)$ from $\int K(y, x)\, f(x)\, dx$, or of solving the integral equation:

$$g(y) = \int_a^b K(y, x)\, f(x)\, dx \qquad [1.1]$$

which happens to be a Fredholm integral equation (because the limits of the integral are fixed, not variable) and of the first kind (because $f(x)$ appears

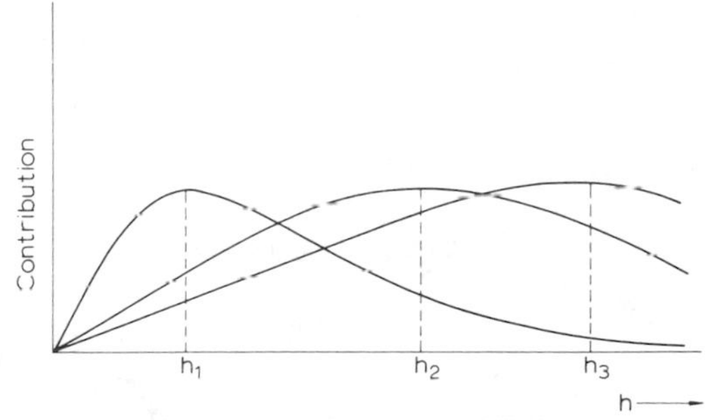

Fig. 1.3. Similar to Fig. 1.2 but representing the situation where the kernels (relative contribution functions) are more spread out.

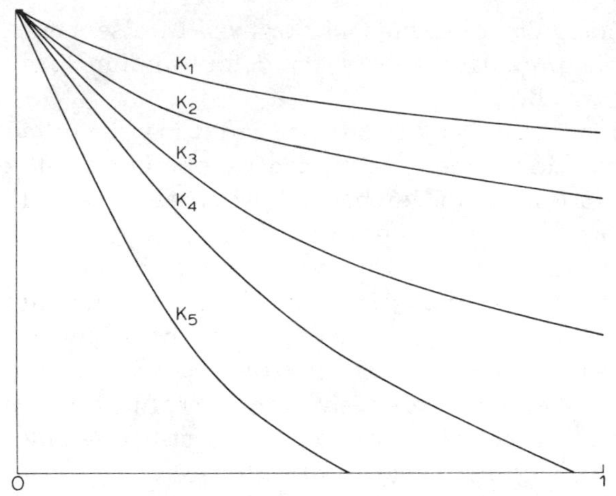

Fig. 1.4. Kernels of the exponential character which decrease from a common maximum at the zero abscissa.

only in the integrand) — the function $K_i(x)$ or $K(\lambda, x)$ is known as the "kernel" or "kernel function".

To the mathematician it will be evident that the inversion problem has some similarities to the problem of inverting Fourier and Laplace transforms. Some physical inversion problems in fact can be reduced to the inversion of the Laplace transform:

$$g(y) = \int_0^\infty e^{-yx} f(x)\, dx \qquad [1.2]$$

One might at first sight believe that all that is needed here is a swift perusal of the appropriate textbook and extraction of a suitable inversion formula. Unfortunately it is not so simple. Here, for example, is a well-known inversion for the Laplace transform:

$$2\pi i f(x) = \int_{\alpha - i\infty}^{\alpha + i\infty} e^{xy} g(y)\, dy \qquad [1.3]$$

the integral to be taken along a path lying to the right of any singularities of $g(y)$.

Another inversion formula, given in Döetsch's (1943) textbook is:

$$f(x) = \lim_{k \to \infty} \frac{(-1)^k}{k!} \left(\frac{k}{x}\right)^{k+1} g^{(k)}\left(\frac{k}{x}\right) \qquad [1.4]$$

In physical inversion problems $g(y)$ is measured only for real values of y, so the data are not able to provide values of $g(y)$ for complex values of y, which [1.3] requires; [1.4] is equally impractical since it requires that we know the functional behavior of $g(y)$ and its derivatives $g^{(k)}(y)$ in the limit $y \to \infty$. This is a region where $g(y)$ vanishes and there is no hope of *measuring* the quantities required for the application of equation [1.4]. Either formula is useful only when the Laplace transform $g(y)$ is given explicitly as a function of y. Experimental measurements of $g(y)$ at many values of y do *not* give g as a function of y in the mathematical sense. Mathematical functions can be evaluated at all values of their arguments — real, imaginary or complex — except where they possess poles or similar singularities. One can fit a functional approximation to a set of measurements of $g(y)$ which will accurately reproduce the measured values, but it is not necessarily true that the approximating function will behave similarly to $g(y)$ for real values of y outside the interval within which the measurements were made. It would be more presumptuous still to use the approximating function with imaginary or complex arguments. This point can be illustrated in a rather simple way by invoking Fourier and Laplace transforms. The Fourier transform of a function $f(x)$ can be written as:

$$F(\omega) = F\{f\} = \int_{-\infty}^{\infty} e^{i\omega x} f(x)\, dx$$

while the Laplace transform is:

$$g(u) = L\{f\} = \int_{0}^{\infty} e^{-ux} f(x)\, dx$$

If $f(x)$ is symmetric about the origin (i.e. if $(f(x) = f(-x))$, then the Fourier and Laplace transforms differ only in that the Fourier transform has an exponential term with the imaginary exponent $i\omega x$ while the corresponding exponent in the Laplace transform is the real number $-ux$. Numerical and analytic methods for inverting Fourier transforms exist and are demonstrably stable and accurate, whereas inversion of the Laplace transform is difficult and unstable. The reason of course is the fundamentally different character of the exponential function for imaginary and for real arguments (Fig. 1.5). In the latter case, the exponential is smooth and monotonic from $x = 0$ to $x = \infty$; in the case of an imaginary argument the exponential becomes oscillatory and its average amplitude is maintained from $x = 0$ to $x = \infty$. It is intuitively obvious that if we wish $\int_0^\infty e^{-ux} f(x)\, dx$ to show some response to values of x around, say, $x = 100$, u must be small enough that e^{-ux} is appreciably different from zero, say $u \sim 0.01$; but the derivative or slope of e^{-ux} is $-u\, e^{-ux}$: if u is small so is the derivative and the function e^{-ux} will change very slowly around $x = 100$ and there will be difficulty in *resolving:* $f(100)$ will be very difficult to separate from $f(99)$ or $f(101)$ and, furthermore, any rapid fluctuation in $f(x)$ around $x = 100$ will give only a

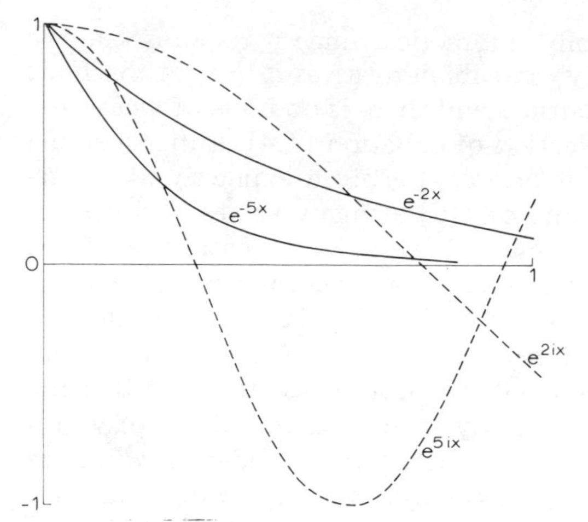

Fig. 1.5. Contrast between the functions e^{-mx} and e^{imx} — kernels for Laplace and Fourier transforms respectively.

very small change in the integral $\int_0^\infty e^{-ux} f(x)\,dx$. This point is illustrated in Fig. 1.6, which shows the very slight variation in $e^{-0.01x}$ between $x = 99$ and 101; the sinusoidal fluctuation shown contributes to the integral a very small amount, which can be calculated to be about 0.001 for a sinusoid of amplitude 1. If we increase u to a value greater than 0.01, the relative slope of e^{-ux} becomes greater but the absolute magnitude of e^{-ux} shrinks and the contribution is still very small. We will return to this point in a subsequent chapter.

If now we look at the behavior of $\int_{-\infty}^\infty e^{i\omega x} f(x)\,dx$ for imaginary exponents $i\omega x$, the situation is very different. $e^{i\omega x}$ oscillates with period $2\pi/\omega$ and amplitude 1 and the magnitude of its derivative or slope does not

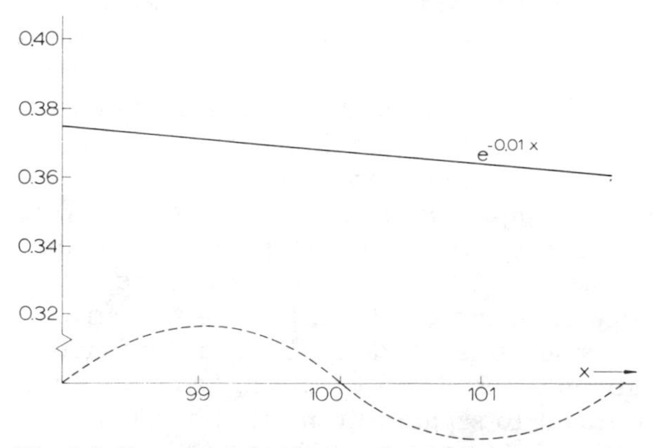

Fig. 1.6. Exponential at large value of the argument x.

decrease monotonically with increasing ω, so there is no loss of resolution as we go out to larger values of x. The amplitude and slope of $e^{i\omega x}$ are the same at 100 as they are at 100 plus or minus any multiple of 2π. For example, at $100 - 31\pi$ or 2.6104, or at $100 - 32\pi$ or -0.5312.

These points show that the Laplace and Fourier transforms, although formally equivalent, are practically very different. They also serve to illustrate a fundamental aspect of inversion problems and of Fredholm integral equations of the first kind: problems with smooth, especially monotonic kernel functions present mathematical and numerical difficulties and their solutions are generally unstable, in the sense that small changes in $g(y)$ can give rise to large changes in the solution. This statement of course is totally equivalent to the statement that smooth and especially monotonic kernels imply integral transforms which are insensitive to certain oscillatory perturbations in $f(x)$ as was depicted in Fig. 1.6. It is this property that presents the fundamental difficulty in inversion problems and it is the avoidance of the resulting instability that is the first requirement for a practical inversion algorithm or numerical procedure.

1.2. EXAMPLES OF REAL INVERSION PROBLEMS

Inference of atmospheric temperature structure

This is now done routinely by means of satellite-borne radiometers, spectrometers and interferometers. The method rests on Planck's Law which gives the rate of emission of radiant energy by a black body at temperature T and wavelength λ:

$$B(\lambda, T) = c_1\lambda^{-5}(e^{c_2/\lambda T} - 1)^{-1}$$

(c_1 and c_2 constants), and Kirchhoff's Law which requires that the absorptivity of a body (or mass of radiating gas) equals its emissivity — i.e. if an element absorbs a fraction α of radiant energy falling upon it, it will emit a fraction α of that which a black body at the same temperature would emit. To determine the balance of monochromatic radiation passing through a layered medium which absorbs and emits — but does not scatter — the radiation, we note that in passing through a vertical distance Δx radiation with intensity $I(\mu)$ will suffer absorption amounting to $\kappa\Delta xI(\mu)/\mu$; μ here is the cosine of the direction of propagation measured from the vertical and κ is the absorption coefficient. But by Kirchhoff's Law the medium will emit thermal radiation amounting to $(\kappa\Delta x/\mu)B$ where B is the Planck radiance at the prevailing temperature T. Thus:

$$\mu\frac{dI(\mu)}{d\tau} = -I(\mu) + B$$

This is one form of the fundamental equation of radiative transfer, in which the optical depth τ is the independent variable, τ being defined by:

$$\Delta\tau = \kappa\Delta x$$

The radiation emerging from the top of the layer is obtained by integrating the radiative transfer equation down to the underlying surface or to a level deep enough for no further measurable contribution to be made to the integral from below. The integral giving the emergent radiation is simply derived from the differential equation since we have:

$$\frac{\mathrm{d}}{\mathrm{d}\tau}(I\,\mathrm{e}^{\tau/\mu}) = \frac{1}{\mu}B\,\mathrm{e}^{\tau/\mu}$$

so if $I = I_0$ at $\tau = 0$, the emerging radiation at $\tau = \tau_0$ is $I(\tau_0)$ if:

$$I(\tau_0)\,\mathrm{e}^{\tau_0/\mu} = \int_0^{\tau_0} B\,\mathrm{e}^{\tau/\mu}\,\mathrm{d}\tau/\mu + I_0$$

or:

$$I(\tau_0) = I_0\,\mathrm{e}^{-\tau_0/\mu} + \int_0^{\tau_0} B\,\mathrm{e}^{-(\tau_0-\tau)/\mu}\,\mathrm{d}\tau/\mu \qquad [1.5]$$

The quantity $\mathrm{e}^{-(\tau_0-\tau)/\mu}$ represents the fraction of radiation passing the level τ which reaches the level of emergence τ_0, thus it represents the *transmission* between the level of emission and emergence. Hence if $t(\tau)$ denotes the transmission between levels τ and τ_0, the intensity emerging is:

$$I = \int_0^1 B\mathrm{d}t(\tau) \qquad [1.6]$$

assuming a layer deep enough for the term $I_0\,\mathrm{e}^{-\tau_0/\mu}$ to be neglected and so effective vanishing of the transmission from $\tau = 0$ to the emergence level $\tau = \tau_0$, so that the emerging radiation all originates within the layer. Equations [1.5] and [1.6] show that the emerging radiance I — which is a function of μ and of wavelength — represents an integral transform of the function B which relates Planck radiance (involving temperature and wavelength) to optical thickness. The optical depth at a given geometric depth is not constant since the absorption coefficient κ and hence the transmission t varies with wavelength. If we measure geometric position by a coordinate x (which may be height, pressure, log pressure or any other quantity, so long as it serves to locate the level unambiguously), then [1.6] can be written:

$$I = \int_{x_0}^{x_1} B(x)\,\frac{\mathrm{d}t(x)}{\mathrm{d}x}\,\mathrm{d}x$$

or:

$$I(\lambda) = \int_{x_0}^{x_1} K(\lambda, x)\, B(x)\, \mathrm{d}x \qquad [1.7]$$

In other words the kernel of the integral equation relating intensity to the distribution $B(x)$ is given by the derivative of the transmission. This integral equation, however, is not linear because $B(x)$ also varies with wavelength. This is usually adjusted by using the Planck function B_0 for some appropriate reference wavelength and temperature and expanding this in a Taylor series of which only the first two terms are retained, i.e. the actual Planck function $B(\lambda_i, T(x))$ for the level x at wavelength λ_i is replaced by its value at a fixed reference wavelength $\bar{\lambda}$, using a linear approximation of the form:

$$B(\lambda, T) = \alpha_\lambda B(\bar{\lambda}, T) + \beta_\lambda$$

At the expense of some loss in accuracy, which is usually not too serious, the non-linear equation [1.7] is replaced by a linear one in which $\bar{B}(x)$ is the Planck function for $\lambda = \bar{\lambda}$, and so a function solely of x, so that an inference of the vertical profile of $\bar{B}(x)$ gives uniquely the vertical temperature profile. The equation to be inverted is:

$$\alpha_\lambda^{-1} \{I(\lambda) - \beta_\lambda\} = g(\lambda) = \int_{x_0}^{x_1} K(\lambda, x)\, \bar{B}(x)\, \mathrm{d}x$$

Fig. 1.7 shows transmittances $\tau(x)$ and kernels $\mathrm{d}\tau(x)/\mathrm{d}x$ for a recent NOAA satellite; x being $(\text{pressure})^{2/7}$ is directed along the *vertical* axis.

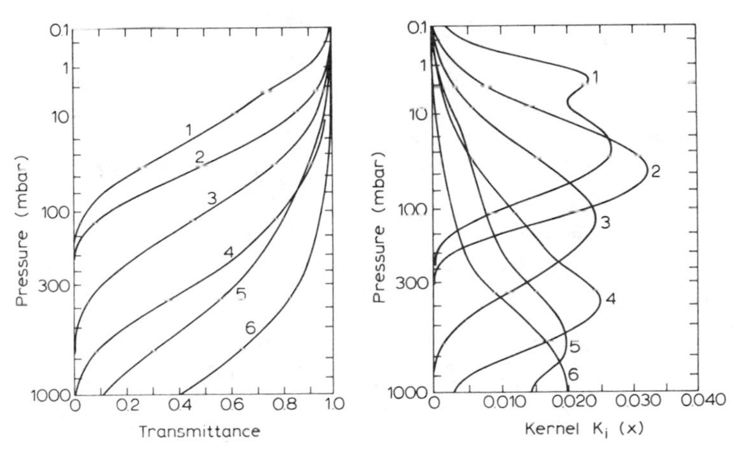

Fig. 1.7. Kernels for a recent satellite temperature-sensing instrument. x is a height measure which is 1 at 0.01 mbar pressure level and 100 at the ground (more strictly 1000 mbars) and varies linearly as $p^{2/7}$.

Measurement of the distribution of an absorbing gas (such as ozone in the earth's atmosphere) from the spectrum of scattered light

In the case of ultraviolet radiation, for example, emission in the atmosphere is totally negligible. Incident radiation $I_0(\lambda)$ is either absorbed or scattered, and if particulate scattering is negligible the scattering from a given elementary volume is proportional to the number of molecules therein, i.e. the layer between pressure levels p and $p + \Delta p$ scatters in proportion to Δp. If there are x units of absorbing gas with absorption coefficient κ above the elementary layer, the scattering by the layer in the direction $\mu = \cos^{-1}\theta$ is:

$$\Delta I_s = S_\lambda(\mu, \mu_0)\, \mathrm{e}^{-\kappa x/\mu_0}$$

here S_λ is a proportionality constant relating scattering power and Δp. The total scattered intensity is obtained by integrating over the entire atmosphere. If the observer is, say, on the ground, where $x = X$ and $p = P$ (by definition), then:

$$I_s = S_\lambda(\mu, \mu_0)\mathrm{e}^{-\kappa X/\mu} \int_0^P \mathrm{e}^{-\kappa x/\mu_0}\, \mathrm{e}^{+\kappa x/\mu}\, \mathrm{d}p$$

$$= S_\lambda(\mu, \mu_0)\mathrm{e}^{-\kappa X/\mu} \int_0^P \mathrm{e}^{-kx}\, \mathrm{d}p \qquad \left[k = \kappa\left(\frac{1}{\mu_0} - \frac{1}{\mu}\right)\right]$$

while if the measurement is from a satellite:

$$I_s = S_\lambda(\mu, \mu_0) \int_0^P \mathrm{e}^{-\kappa x/\mu_0}\, \mathrm{e}^{-\kappa x/\mu}\, \mathrm{d}p$$

$$= S_\lambda(\mu, \mu_0) \int_0^P \mathrm{e}^{-kx}\, \mathrm{d}p \qquad \left[k = \kappa\left(\frac{1}{\mu_0} + \frac{1}{\mu}\right)\right]$$

k can be varied by varying the wavelength (since κ is wavelength-dependent), by varying the direction of view μ, or by allowing μ_0, the incident solar illumination's direction, to vary (as the sun rises or sets). In all situations the mathematics of the problem is the same: given a set of measurements of $\int_0^P \mathrm{e}^{-kx}\, \mathrm{d}p$ for $k = k_1, k_2, \ldots, k_m$, to infer the distribution of $x(p)$, or $p(x)$. Integration by parts gives:

$$\int_0^P \mathrm{e}^{-kx}\, \mathrm{d}p = \int_0^X \mathrm{e}^{-kx} \frac{\mathrm{d}p}{\mathrm{d}x}\, \mathrm{d}x = \left|\mathrm{e}^{-kx}\, p(x)\right|_0^X + k \int_0^X \mathrm{e}^{-kx}\, p(x)\, \mathrm{d}x$$

$$= \mathrm{e}^{-kX} P - p(0) + k \int_0^X \mathrm{e}^{-kx}\, p(x)\, \mathrm{d}x$$

Usually P is known; X, the total depth of absorber, can be determined and p_0 set to zero, in which case we have a linear integral equation:

$$g(k) = \int_0^X e^{-kx}\, p(x)\, dx \qquad [1.8]$$

the kernel of which is the exponential function e^{-kx}. In the previous example also the kernel — the transmission function — was exponential (for monochromatic radiation) or a combination of exponentials (for finite wavelength intervals).

Umkehr. This word implying "reversal" or "turning about" was used to describe the behavior of the ratio of the solar energy measured at the ground at two wavelengths, one being more strongly absorbed by atmospheric ozone than the other. This ratio decreases as the sun drops towards the horizon but at low solar angles its trend is reversed and it commences to increase again.

A very simplified model of the atmosphere helps to explain this result. Consider a layer of the atmosphere to lie above the ozone and the rest of the atmosphere to be below it, there being n_1 molecules in the first layer and n_2 in the second. If there are x units of ozone and the absorption coefficient is k, the signal received by a detector at the ground is proportional to:

$$e^{-kx}\, n_1 + e^{-kx \sec\theta}\, n_2$$

where θ = solar zenith angle).

If the measurement is made at a second wavelength where the absorption coefficient is k' then:

$$\frac{I}{I'} = \frac{e^{-kx}\, n_1 + e^{-kx\sec\theta}\, n_2}{e^{-k'x}\, n_1 + e^{-k'x\sec\theta}\, n_2} = e^{-(k-k')x} \frac{n_1 + n_2\, e^{-kx(\sec\theta - 1)}}{n_1 + n_2\, e^{-k'x(\sec\theta-1)}}$$

The variation of the ratio (I/I') with θ is therefore governed by:

$$\frac{1 + (n_2/n_1)\, e^{-kx(\sec\theta-1)}}{1 + (n_2/n_1)\, e^{-k'x(\sec\theta-1)}}$$

for $kx(\sec\theta - 1)$ small and $n_2 >> n_1$, the ratio is dominated by the second terms and an exponential decrease in the ratio occurs when $k > k'$. If, however, $n_2\, e^{-kx(\sec\theta-1)}$ becomes small compared with n_1, while $n_2\, e^{-k'x(\sec\theta-1)}$ is still *not* negligible compared to n_1, the ratio is approximately:

$$\frac{n_1}{n_1 + n_2\, e^{-k'x(\sec\theta-1)}}$$

and in this region an *increase* in the ratio occurs. Regular Umkehr data have been obtained at a number of observatories in considerable detail for many years. The inversion of Umkehr data to obtain a distribution curve is analogous to the inversion of spectral data obtained for a fixed solar angle (since in one case the absorption coefficient is varied, in the other it is the solar angle). The actual ratio I/I' is not linearly related to ozone distribution or the equivalent $p(x)$ of equation [1.8], unless I' is not absorbed at all. The ratio can be written (using the same simplifying approximation used in equation [1.8]):

$$r(\theta) = \int_0^x e^{-kx}\, dp \Big/ \int_0^x e^{-k'x}\, dp$$

Inference of aerosol particle size distribution from measurements of the fraction of the total number transmitted through filters or diffusion batteries

Filters or diffusion batteries (parallel-plate arrangements with close spacing between the plates) remove particles as a result of diffusion of particles to the walls of the filter pores or of the plates. In filters larger particles are impacted on the front surface of the filter and very large particles may even fall out under gravity. Each of these removal mechanisms can be analyzed mathematically and when the particles are spherical and their density known, it is possible to calculate the fraction removed for a given radius and a given flow. This fraction changes with flow rate and with geometry for a given sized particle and varies with particle size for a fixed flow rate and geometry. If an aerosol containing an unknown size distribution is passed through a filter and the fraction of the particles which emerge from the filter is measured, we find that this fraction changes as the flow rate is varied; if another different filter is substituted, a different relationship of fraction transmitted to flow rate is found. In fact, if $K_i(x)$ represents the filter transmission as a function of particle size x for the ith measurement, then the fraction emerging is:

$$g_i = \int_0^\infty K_i(x)\, f(x)\, dx$$

if $f(x)\,\Delta x$ is the fraction of the particles in the size interval $x \to x + \Delta x$. Thus once again a measured quantity g is related to the unknown distribution by a linear integral relationship.

Dissipation of seismic waves

The earth can oscillate mechanically in an infinity of normal modes. From seismic records the amplitude and damping of these normal modes (which can be separated by frequency-filtering the original time series) can be measured. If the fraction of the energy in the ith mode which is dissipated per cycle is g_i, and $K_i(r)$ is the ratio of the elastic shear energy in the shell $r \to r +$

Δr to the total shear energy in mode i, then, so far as first-order theory is applicable:

$$g_i = \int_0^1 K_i(r)\, f(r)\, dr$$

if $f(r)$ is the fraction of the energy density at $r \to r + \Delta r$ which is dissipated as heat in one cycle. This dissipation function is related to the composition and density of the earth which, of course, cannot be measured directly — again the integral relationship connects the inaccessible function being sought to the measurable quantities g_i. As in most physical inversion problems the physical nature of $f(r)$ dictates that it be non-negative and as in most inversion problems the kernels also are non-negative.

Decay of mixtures of radioactive materials

If a units of a radioactive material which decays with a time constant α, b units of a radioactive material with time constant β, and so on, are decaying together as a mixture, the concentration of each is $a\, e^{-\alpha t}$, $b\, e^{-\beta t}$, etc. The rate of decay of the mixture at time t is:

$$g(t) = a\alpha\, e^{-\alpha t} + b\beta\, e^{-\beta t} + \ldots$$

or, if we adopt a subscript notation:

$$g(t) = \sum_j a_j \alpha_i\, e^{-\alpha_j t}$$

It is possible to describe the situation by considering the time constant α, a variable categorizing a particular component, and $f(\alpha)$, a distribution describing the composition, by specifying that there were $f(\alpha)\, \Delta\alpha$ units originally present with time constant between α and $\alpha + \Delta\alpha$, so that:

$$g(t) = \int_0^\infty \alpha\, e^{-\alpha t} f(\alpha)\, d\alpha$$

This equation shows that $g(t)$ is the Laplace transform of the function $\alpha f(\alpha)$, and again determination of $f(\alpha)$ requires that an integral equation be inverted. The above equation is the same as that encountered in the atmospheric ozone problem discussed earlier on pp. 12—13.

Aperture synthesis

A plane wave front coming to a mirror, antenna or similar receiving surface from the direction (θ, ϕ) arrives at different parts of the receiving surface with different phases; if the contributions from all the elements making up the reflector are combined electrically before detection the resulting signal is obtained by integration over all area elements, taking the phase into account, so that the integration must be made over complex amplitude. An element δA in the horizontal plane at a distance r from the origin and azimuthal angle ϕ' from the azimuth reference direction receives a signal with a phase $(2\pi r/\lambda) \sin\theta \cos(\phi' - \phi)$ relative to the phase at the coordinate origin. The wavelength λ can be set equal to unity without loss of generality and one obtains from the element δA a contribution to the signal which is proportional to:

$$e^{2\pi i r \sin\theta \cos(\phi' - \phi)} \delta r \delta\phi'$$

for a flat horizontal receiving surface the signal is obtained by integrating the above over the entire surface. For a *ring* of radius a the result is:

$$f(\theta, \phi) = \int_0^{2\pi} e^{2\pi i \sin\theta \cos u} \, du = J_0(2\pi a \sin\theta)$$

using Bessel's integral:

$$J_k(z) = \frac{1}{2\pi} \int_{-\pi}^{\pi} e^{iku - iz \sin u} \, du$$

If the entire circular aperture was filled in, the signal for a given input would be increased (in proportion to the area), but the angular resolution is not appreciably different from that of the ring; the amplitude response of the disk is proportional to:

$$\frac{J_1(2\pi a \sin\theta)}{2\pi a \sin\theta}$$

(The first zero of $x^{-1} J_1(x)$ occurs at 3.8317, the first zero of $J_0(x)$ at 2.4048, so the resolution is actually slightly better for the ring, but the sidelobes are greater). It is the linear dimension a which determines the angular resolution. The ring, however, exhibits much larger sidelobes (maxima and minima in the complex response) which lead to an undesirable uncertainty in locating a distant source, since a source on the axis cannot be distinguished from one at the other maxima and minima of $J_0(2\pi a \sin\theta)$ — the negative minima of $J_0(2\pi a \sin\theta)$ represent maxima in the intensity, since intensity is given by $|\text{amplitude}|^2$.

The process of aperture synthesis consists of combining the signals from members of a system of small receiving elements so as to emulate a much larger single receiver (which would be prohibitively expensive and perhaps technologically unattainable), so as to obtain a resolution comparable to that of the latter (sensitivity cannot be made comparable since it depends on area). If the complex polar diagram of the nth receiving element is $f_n(\theta, \phi)$ the mathematical process of aperture synthesis amounts to determining complex coefficients c_n so as to optimize the combined response. One could for example match $\Sigma_{n=1}^{N} c_n f_n(\theta, \phi)$ to obtain a precise agreement with an arbitrary system response $R(\theta, \phi)$ at $2N$ directions (θ_1, ϕ_1), (θ_2, ϕ_2), etc. However, the requirements of symmetry reduce the number of disposable variables greatly. If the N receiving elements are positioned at uniform spacing around a circle, the coefficients must be equal in modulus. Furthermore, for circular symmetry the possible phase changes to be applied to the elements must give a total phase change around the annulus of 2π or a multiple thereof.

Wild (1965) has discussed and constructed a large annular array of paraboloidal radio reflectors in which the image is formed by aperture synthesis in which the c_n are $\exp^{(2\pi ik\phi_n')}$, the outputs of the interconnected elements being combined coherently for various values of the integer k before detection (which gives a signal proportional to $|\text{amplitude}|^2$). Considering the summation over the discrete finite elements as equivalent to integration over an infinity of continuous infinitesimal elements, the amplitude response is found to be proportional to $J_k(2\pi a \sin\theta)$ (because of the additional $e^{ik\phi'}$ term as compared to the annulus). When the combined signal is detected the result is an output proportional to $J_k^2(2\pi a \sin\theta)$. Wild has derived analytic formulae whereby a spatially band-limited set of symmetric functions can be synthesized by a sum:

$$R(\theta) = \sum_k t_k J_k^2(2\pi a \sin\theta) \qquad [1.9]$$

It is clear that if the t_k were allowed to become too large, instability and gross error magnification would result. The problem of determining a set of t_k to obtain a desired behavior of $R(\theta)$ is essentially an inversion problem. An alternative approach to Wild's analytic method would be the determination of a set of t_k with bounded norm (so as to optimize a suitable measure of the spread of $R(\theta)$). The quantity $\int \theta^2 R^2(\theta) d\theta$ could for example be used, the solution t_k ($k = 1, 2, ..., N$) being derived by the methods discussed in Chapter 7.

Inference of aerosol- or drop-size distributions from optical transmission or scattering measurements

The attenuation of light of wavelength λ passing through air containing spherical aerosol particles or drops is given by the equation:

$$\frac{I}{I_0} = \exp\{-l \int_0^\infty \pi r^2 Q_e(r/\lambda)\, n(r)\, dr\}$$

I_0 being the incident irradiance or illuminance and I the irradiance or illuminance after traversal of a path of length l through a population containing $n(r)\ \Delta r$ drops of particles per unit volume. $Q_e(r/\lambda)$, the extinction efficiency, can be calculated from Mie scattering formulae provided the refractive index is known. Fig. 1.8 shows how $Q(r/\lambda)$ varies with r/λ for refractive index 1.33. If the ratio I/I_0 is determined experimentally over a range of wavelengths and if $\pi^{-1}\ l^{-1} \log I/I_0$ (which varies with wavelength) is defined to be $g(\lambda)$, then evidently:

$$g(\lambda) = \int r^2 Q(r/\lambda)\ n(r)\ \mathrm{d}r \qquad [1.10]$$

and once again a linear integral equation of the first kind relates the unknown $n(r)$ with the measured $g(\lambda)$, the kernel being $r^2 Q(r/\lambda)$. This is also plotted in Fig. 1.8.

In a completely analogous way light scattered at different scattering angles may be measured at one wavelength, or different wavelengths may be used combined with one or more angles, to provide data for the inference of size distribution by inversion. The kernels in this case are the Mie scattering functions usually written as i_1 and i_2, which are functions of angle, wavelength, particle size and refractive index. If unpolarized light is employed, $(i_1 + i_2)$ becomes the kernel, but other arrangements involving polarization are of course possible. Such measurements may be made in samples brought into

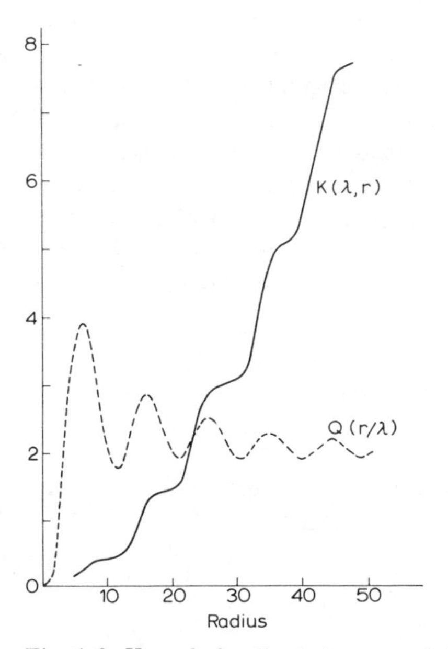

Fig. 1.8. Kernels for the inference of size distribution from optical transmission.

an instrument or they may be made remotely, as in so-called "bistatic lidar" where atmospheric scattering at different scattering angles is measured by varying the direction of an illuminating optical system and a receiving system. The intersection of the two "beams" can be kept at a constant height level but with such a bistatic system different parts of the atmosphere are viewed for different scattering angles and any horizontal inhomogeneity of the atmosphere will give rise to errors. Irrespective of whether transmitted or scattered radiation is measured, any of the above techniques, or any combination thereof, can be described by linear integral equations, provided only that the optical thickness of the illuminated layer is kept well below unity (if it is comparable to or larger than unity, multiple scattering becomes appreciable and must somehow be taken into account). In other words the measured intensity can be written:

$$I(\lambda) = \int_0^\infty K(\lambda, r)\, n(r)\, \mathrm{d}r \qquad [1.11]$$

for the scattered-light measurements, with $K(\lambda, r)$ given by the appropriate scattering diagram of a particle of radius r at wavelength λ. The same equation holds for $\log\,[I(\lambda)/I_0(\lambda)]$ in the case of measurements of transmitted light, but $K(\lambda, r)$ now represents $r^2 Q_e(r/\lambda)$.

Other optical parameters of scattered light — for example, the polarization ratio — are sensitive to particle size and experimentally are sometimes easier to measure accurately. However, they may follow relationships which are non-linear with respect to $n(r)$; for example, the ratio of scattered light polarized transverse to that polarized parallel to the plane of scattering is given by:

$$R(\theta, \lambda) = \frac{\int i_1(r/\lambda, \theta)\, n(r)\, \mathrm{d}r}{\int i_2(r/\lambda, \theta)\, n(r)\, \mathrm{d}r} \qquad [1.12]$$

Comparatively little work has been done on the inversion of non-linear relationships such as this. They may be amenable to methods similar to the non-linear iterative techniques which are described in Chapter 7.

Path length distribution in radiative transfer

It is possible by doubling and similar superposition techniques to derive the scattering functions of a layer, possibly optically deep (it may even be infinitely deep). This can be done without ever following the detailed paths taken by individual photons, but in some problems it is important to know how these path lengths are distributed. The scattering functions — transmission and reflection functions — depend on incident and emerging angle, optical thickness τ of the layer, single scattering albedo ϖ_0 and the scattering diagram or phase function which gives the distribution of scattering angle, i.e. change of direction after scattering, for single scattering events. The single scattering albedo $\overline{\omega_0}$ gives the fraction of the radiation removed from its former direction of propagation (say in traversing an element of thickness Δz or optical thickness $\Delta\tau$) which reappears after scattering as radiation in other directions of propagation; $\overline{\omega_0} = 1$ means conservative lossless scattering, while $\overline{\omega_0} = 0$ implies absorption without scattering. Suppose one knows the reflection function of a layer as a function of $\overline{\omega_0}$; writing this as $R(\overline{\omega_0})$ — omitting for simplicity explicit reference to the other dependences of R (on τ and on the geometrical variables) — one can put:

$$R(1) = \int_0^\infty p(l)\,\mathrm{d}l$$

where $p(l)\ \Delta l$ represents the fraction of reflected photons which travelled optical paths between l and Δl between entering the layer and emerging. If absorption is now taken into account, it requires an additional exponential term which can be written $\mathrm{e}^{-(1-\overline{\omega_0})l}$, since $(1-\overline{\omega_0})$ is the fraction of the energy removed which is actually absorbed while Δl is the energy removed, by definition of optical path. Absorption within the particles can change the scattering diagram, but in many situations this effect can be legitimately neglected and, of course, in many situations absorption takes place only in the medium between the scatterers while the actual scatterings are conservative; in either event one can write the emerging intensity in the presence of absorption in terms of the distribution of path lengths in the absence of absorption:

$$R(\overline{\omega_0}) = \int_0^\infty \mathrm{e}^{-(1-\overline{\omega_0})l}\,p(l)\,\mathrm{d}l = \int_0^\infty \exp\left(-\frac{1-\overline{w_0}}{\overline{w_0}}\right) p_\mathrm{s}(x)\,\mathrm{d}x \qquad [1.13]$$

and again a linear integral equation of the first kind is obtained relating the known $R(\overline{\omega_0})$ with the unknown $p_\mathrm{s}(x)$. This relationship of course is a Laplace transform, with $(1-\overline{\omega_0})$ the variable of the transformation. The problem is therefore formally equivalent to that discussed on pp. 11—12.

Travel time and velocity structure in seismology

A knowledge of the propagation speed of seismic waves is of great value in geophysics in that these speeds are determined by the density and elastic constants of the medium, which can be determined from laboratory studies for various rock types, etc.

If an artificial earthquake (e.g. an explosion) takes place at a point F at or beneath the earth's surface, its influence will be felt at distant points and at such a distant point A this influence will extend over a period of time even if the original disturbance is very brief. This is, of course, because there are several modes of oscillation and several possible paths; the first waves to arrive at A for stations within 100° or so of the disturbance are compressive waves (P waves) analogous to sound waves, which have travelled a moderately curved path from F to A but have not penetrated to the base of the earth's mantle. The travel time between F and A is an obviously important quantity for locating the focus of an earthquake and extensive tables of travel times have been drawn up for this purpose. The travel time also represents a kind of sounding of the earth's crust and mantle because it is related to the vertical distribution of propagation speed (i.e. the "velocity structure") and hence of composition. This aspect of travel time has been exploited in exploration and in basic studies of earth structure. The propagation speed of seismic waves generally increases with depth below the surface and as a result downward-travelling ray paths are bent (analogously to light rays travelling through a medium in which refractive index is varying) turning increasingly towards the horizontal, eventually turning upwards and reaching the surface. If a ray starts at F and returns to the surface at P, taking a time T to do so, the refraction of the ray in a spherically symmetric earth is such that $p = (r \sin \alpha)/v$ remains constant throughout (r being the radial distance from the centre of the earth, v the speed and α the angle of travel, measured from the direction of the radius vector). The angle Δ subtended at the center of the earth by PF is related to be travel time T by the simple relationship $\mathrm{d}T/\mathrm{d}\Delta = p$ [we are following the notation of Bullen (1963) and the reader is referred to that text for details]. If the travel time of a disturbance is measured at a number of stations, the parameter p for the wave arriving at each station can therefore be estimated by numerical differentiation. In this way either the travel time T or Δ can be plotted against p. It can furthermore be shown that the explicit relation between Δ and p is:

$$\Delta(p) = \int_{\eta_p}^{\eta_0} 2pr^{-1}(\eta^2 - p^2)^{-1/2} \frac{\mathrm{d}r}{\mathrm{d}\eta}\,\mathrm{d}\eta = \int_{\eta_p}^{\eta_0} K(p, \eta)\, f(\eta)\,\mathrm{d}\eta \qquad [1.14]$$

where η is written for r/v; the subscript zero refers to the surface and the subscript p to the lowest point along the ray path. Observational data of T versus Δ can be recast as data for Δ versus p and we arrive thereby at an integral equation relating $[r^{-1}(\mathrm{d}r/\mathrm{d}\eta)]$, viewed as an unknown function $f(\eta)$ of η, to values (derived from measurements) of $\Delta(p)$. The kernel of the integral equation is $2p(\eta^2 - p^2)^{-1/2}$. If the integral equation can be solved for $r^{-1}(\mathrm{d}r/\mathrm{d}\eta)$ as a function of η, it is a trivial matter to obtain from that $v = r^{-1}\eta$ as a function of r.

The integral equation just discussed differs from those obtained in previous sections in one important respect — one of the integration limit, η_p, itself varies with p. This kind of integral equation is more stable and more readily soluble than those with fixed limits, for when we vary η_p, say, from b

to $b + \delta b$. $\Delta(p)$ receives a contribution from the interval $b \to b + \delta b$, which had no influence on previous values of Δ (i.e. values of Δ for $\eta_p \leqslant b$), so new strips of the graph of f against η are being continually revealed as we increase the width of the integration interval. Such integral equations are almost always amenable to iterative solution methods.

The above integral equation even has an analytic solution due to G. Rasch which is derived in Bullen's (1963) textbook. This solution is:

$$\int_0^{\Delta_1} \cosh^{-1}(p/\eta_1)\,\mathrm{d}\Delta = \pi \ln(r_0/r_1) \tag{1.15}$$

A second approach to the problem of inferring velocity structure (Bullen, (1963), section 12.1.2) considers a model crust consisting of m homogeneous horizontal layers, with an underlying medium associated with index $m + 1$. For a near disturbance for which penetration into the earth is slight the delay time for a surface source is approximately:

$$T = \frac{r_0\Delta}{v_p} + 2\int_0^{z_p} (v^{-2} - v_p^{-2})^{1/2}\,\mathrm{d}z = \frac{r_0\Delta}{v_p} + a_p \tag{1.16}$$

From data for a network of stations it is possible to determine v_p and a_p by least squares (note that the second term a_p is independent of station-to-source distance Δ, being determined wholly by the velocity structure down to depth z_p). If the unknown thickness of the jth layer is H_j, then a_p can be expressed as:

$$a_p = 2\sum_{j=1}^{p-1} \alpha_{pj} H_j \qquad (p = 2, 3, \ldots, m + 1) \tag{1.17}$$

α_{pj} being $(v_j^{-2} - v_p^{-2})^{1/2}$. When the latter quantities have been estimated this last equation becomes a system of linear equations for the unknowns H_1, H_2, ..., H_m. Such systems of equations will be discussed in the following chapters.

Inference of earth conductivity with depth from surface measurement of apparent conductivity

Consider two parallel conductors running along the surface of the ground, separated by a distance x. When a potential difference V is applied between the conductors the resulting field distribution below the surface can be calculated readily if the permittivity is constant or varies in a known way.

Along the plane of symmetry, the field $\boldsymbol{E}$ must be horizontal and is a function of depth z. Changing the separation distance x will evidently modify the field and it should therefore be written $E_0(x, z)$. If the conductivity at depth z is $\sigma(z)$ the current at electrode separation x is related to the latter according to the integral equation

$$i(x) = \int_0^\infty E_0(x, z)\, \sigma(z)\, dz \qquad [1.18]$$

By varying the separation distance and measuring the current as a function of x, the vertical distribution of conductivity can be inferred if the integral equation [1.18] can be inverted. Again the integral equation is a linear Fredholm equation of the first kind; again the kernel is a positive quantity which necessarily decreases monotonically with x.

Practically used electrode configurations are generally more elaborate than those envisaged above, but the principles are the same. Variations of dielectric constant with depth and deviations from symmetry necessarily complicate the interpretation of such measurements and, of course, result in equations always containing an appreciable error component.

Downward continuation in potential theory

Considering the earth's surface as an x, y plane and an inwardly-directed z-axis, a fundamental problem in potential theory (Dirichlet's problem) can be stated as follows: When $f_0(x, y)$ is the value of a function $f(x, y, z)$ on the boundary plane $z = 0$, to find a solution of Laplace's equation $\nabla^2 f = 0$ which is continuous on and above the boundary plane, regular in the half space $z \leqslant 0$ and which coincides with $f_0(x, y)$ on the boundary plane.

It can be proved that such a solution exists and that it is unique. Reference to standard textbooks such as Kellogg (1960) gives for the solution $f(x, y, z)$ in the half space:

$$f(x, y, z) = \frac{z}{2\pi} \int_{-\infty}^{\infty} \int_{-\infty}^{\infty} \frac{f_0(\xi, \eta)}{[(x-\xi)^2 + (y-\eta)^2 + z^2]^{3/2}}\, d\xi d\eta \qquad [1.19]$$

In practical situations the infinite integrals can be replaced by integrals over finite limits with negligible loss in accuracy, since the kernel falls off with (distance)$^{-3}$.

Equation [1.19] differs from those encountered in previous examples only in that it involves a double integral rather than a single integral. It is also to be noted that the kernel is a function of the differences $(x-\xi)$ and $(y-\eta)$; this kind of kernel is known as a convolution kernel and the integral is a convolution integral. Convolution represents a kind of smearing of the function $f_0(\xi, \eta)$ analogous to what an optical spectrum is subjected to when it is viewed through the slit of a spectrometer. The function $f(x, y, z)$ produced by the convolution is necessarily a smoother function of x and y than the function $f_0(x, y)$. The smaller the value of z the sharper the peak of the kernel function at $\xi = x$, $\eta = y$ and hence the smaller the smoothing effect of the convolution.

The process of calculating $f(x, y, z)$ for some value of $z(\leqslant 0)$, given $f_0(x, y)$, is a straightforward, stable computation — the kernel is everywhere non-negative and free of rapid oscillations. For numerical evaluation the gravity effect is first usually averaged over a circle of radius r centered at (x, y); writing:

$$h(x, y, r) = \frac{1}{2\pi} \int_0^{2\pi} f_0(x - r\cos\theta, y - r\sin\theta)\, d\theta$$

to represent such an average, relation [1.19] can be expressed as:

$$f(x, y, z) = |z| \int_0^{\infty} h(x, y, r)(z^2 + r^2)^{-3/2}\, r dr \qquad [1.20]$$

which involves a single integral and can be evaluated by a suitable weighted sum (i.e. a quadrature).

The process of obtaining $f(x, y, z)$ from $f_0(x, y)$ is known for obvious reasons as "upward continuation". The inverse process, that of obtaining $f_0(x, y)$ from $f(x, y, z)$ is "downward continuation". Below the earth's surface Poisson's equation (rather than Laplace's equation) applies, but if the regular part of the gravity vector including centrifugal and tidal influences are first subtracted out, local and often minor deviations of the gravity field remain which are ascribed to subsurface masses. The vertical component of the gravity field anomaly describes the heterogeneity of the upper part of the earth's crust. If we look only at the anomalies or "gravity effects", downward continuation would enable disturbing masses to be located and their area and magnitude estimated. With no disturbing masses above the level z, the gravity effect at z can be obtained in principle by inversion of [1.19] or [1.20], treating the level z as the reference level and viewing the integral relationship as giving the (known) surface values by integral operation on the now unknown $f_0(x, y)$.

BIBLIOGRAPHY

Bullard, E.C. and Cooper, R.I.B., 1948. The determination of the masses necessary to produce a given gravitational field. *Proc. R. Soc. Lond., Ser. A*, 194: 332—347.

Bullen, K.E., 1963. *Introduction to the Theory of Seismology*. Cambridge University Press, Cambridge, 381 pp.

Döetsch, G., 1943. *Theory and Application of the Laplace Transformation*. Dover, New York, N.Y., 439 pp. (translation).

Goetz, F.W., 1951. Ozone in the atmosphere. In: T.F. Malone (Editor), *Compendium of Meteorology*. American Meteorological Society, Boston, Mass., 1334 pp.

Grant, F.S. and West, G.F., 1965. *Interpretation Theory in Applied Geophysics.* McGraw-Hill, New York, N.Y., 583 pp.
Kaplan, L.D., 1959. Inference of atmospheric structure from remote radiation measurements. *J. Opt. Soc. Am.*, 49: 1004—1006.
Kellogg, O.D., 1960. *Foundations of Potential Theory.* Dover, New York, N.Y., 384 pp.
Wark, D.Q. and Fleming, H.E., 1966. The determination of atmospheric temperature profiles from satellites, I. Introduction. *Mon. Weather Rev.*, 94: 351—362.
Wild, J.P. 1965. A new method of image formation with annular apertures and an application in radioastronomy. *Proc. R. Soc. Lond., Ser. A*, 286: 499—509.

CHAPTER 2

Simple problems involving inversion

Over the years there have been many brain-teasers similar to the following:

"A zoo-keeper, when asked how many birds and animals were in his care, replied, 'I have one hundred head, three hundred feet' How many of each did he indeed have in his care?"

It is not difficult to assign the minimum of two feet to each of the heads; noting that the additional fifty pairs of feet must be supplied by animal members of the zoo population, one readily arrives at a count of fifty animals and fifty birds.

Yet this elementary puzzle contains the essential ingredients of all inversion problems in that the data supplied do not separately list the numbers in the possible categories, but rather gives the values of linear combinations of these. In other words if a is the number of animals and b the numbers of birds, the data supplied by the keeper were:

$$a + b = 100 \text{ (one head per animal, one per bird)}$$
$$4a + 2b = 300 \text{ (four feet per animal, two per bird)}$$

in other words two equations to determine two unknowns. As all students of elementary algebra know, that suffices for the determination of two unknowns. . . but not always.

Suppose now the zoo contained only animals, lions and tigers say. If the keeper again reported 100 head, 300 feet, then with l lions and t tigers the equations would be:

$$l + t = 100$$
$$4l + 4t = 300$$

These equations are contradictory, for the second is equivalent to $l + t = 75$, which the first states to be 100. (The keeper, of course, was lying — there is no way 100 head and 300 feet can occur simultaneously.) If on the other hand the keeper had 50 of each species and replied truthfully that he had 100 head, 400 feet, the equations would be:

$$l + t = 100$$
$$4l + 4t = 400$$

These equations are not contradictory, but do not contain enough information to separate the values of l ant t.

Thus it is not always true that N equations are sufficient to give N unknowns: the equations may be mutually contradictory or they may be consistent but contain insufficient information to separate N unknowns.

When there are a small number of equations and unknowns, simple inspection will often be enough to indicate which situation we are confronted with, especially when the coefficients are comparable and whole numbers. In any event, the method of algebraic elimination provides a simple means for separating the unknowns.

2.1. ALGEBRAIC ELIMINATION

This procedure depends on multiplication (or division) of individual equations by judiciously chosen numbers, followed by subtraction (or addition) of pairs of equations so as to eliminate one unknown; the process is repeated until an equation in one unknown only is obtained. This unknown thereby becomes known, and the other unknowns follow readily by substitution. Consider, for example, the equations:

$$x + y + z = 2 \quad [i]$$

$$2x - y + 3z = 9 \quad [ii]$$

$$3x + 2y - z = -1 \quad [iii]$$

If we multiply [i] by 2 and subtract from [ii] we obtain:

$$0 \cdot x - 3y + z = 5 \quad [iv]$$

while multiplied by 3 and subtracted from [iii] it gives:

$$0 \cdot x - y - 4z = -7 \quad [v]$$

The two equations [iv] and [v] in the two unknowns y and z are reduced to an equation in z alone if y is eliminated by multiplying [v] by 3 and subtracting [iv]. Thus:

$$-13z = -26 \quad [vi]$$

So:

$$z = 2$$

Insert this into [iv]:

$$-3y + 2 = 5, \text{ or } y = -1$$

and x, by equation [i], is found to be 1. The solution process is then completed by verifying that when we set $x = 1$, $y = -1$, $z = 2$, [i], [ii] and [iii] become identities:

$$x + y + z = 1 - 1 + 2 = 2$$

$$2x - y + 3z = 2 + 1 + 6 = 9$$

$$3x + 2y - z = 3 - 2 - 2 = -1$$

Thus a complete and painless process for solving the equations exists. Consider next the equations:

$$x + y + z = 2 \qquad [a]$$

$$2x - y + 3z = 9 \qquad [b]$$

$$4x + y + 5z = 12 \qquad [c]$$

y is most readily eliminated, by addition of [a] and [b] and of [c] and [b]:

$$3x + 4z = 11$$

$$6x + 8z = 21$$

These are contradictory (the left side of the second equation is exactly twice the left side of the first and if we attempt to eliminate x by multiplying the first equation by 2 and subtracting from the second, we obtain:

$$0 = -1(!)$$

The theory of linear algebraic equations shows that the problems appear when the determinant [1] of the coefficients vanish. In the example just given, this determinant is:

$$\Delta = \begin{vmatrix} 1 & 1 & 1 \\ 2 & -1 & 3 \\ 4 & 1 & 5 \end{vmatrix}$$

or, expanding in minors along the top row:

$$\begin{vmatrix} -1 & 3 \\ 1 & 5 \end{vmatrix} - \begin{vmatrix} 2 & 3 \\ 4 & 5 \end{vmatrix} + \begin{vmatrix} 2 & -1 \\ 4 & 1 \end{vmatrix}$$

which equals:

$$-8 - (-2) + 6 = 0$$

One might then conclude that if we first evaluate the determinant Δ to make sure that the troublesome condition $\Delta = 0$ (known as "singularity") does not exist, then continued elimination in some form must give the solution, regardless of how many unknowns are involved. When there are

[1] See Appendix for definition and elementary properties of determinants,

many unknowns many elimination steps are needed, but the whole process can readily be carried out by computer. In the general case:

$$\left.\begin{array}{l} a_{11}x_1 + a_{12}x_2 + \ldots + a_{1N}x_N = y_1 \\ a_{21}x_1 + a_{22}x_2 + \ldots + a_{2N}x_N = y_2 \\ \cdot \\ \cdot \\ a_{N1}x_1 + a_{N2}x_2 + \ldots + a_{NN}x_N = y_N \end{array}\right\} \quad [2.1]$$

a computer program could eliminate x_1 by division of all member equations by the coefficient of x_1, so that the following set of equations is formed:

$$\left.\begin{array}{l} x_1 + a'_{12}x_2 + a'_{13}x_3 + \ldots + a'_{1N}x_N = y'_1 \\ x_1 + a'_{22}x_2 + a'_{23}x_3 + \ldots + a'_{2N}x_N = y'_2 \\ \cdot \\ \cdot \\ x_1 + a'_{N2}x_2 + a'_{N3}x_3 + \ldots + a'_{NN}x_N = y'_N \end{array}\right\}$$

here a'_{jk} is written for a_{jk}/a_{j1}. Subtraction of any of these equations from each of the others yields a system of equations with one less variable — i.e. $N - 1$ equations in $N - 1$ unknowns. If, for example, the last equation is subtracted from all the others, we have $N - 1$ new equations:

$$\left.\begin{array}{l} (a'_{12} - a'_{N2})\, x_2 + (a'_{13} - a'_{N3})\, x_3 + \ldots = y'_1 - y'_N \\ (a'_{22} - a'_{N2})\, x_2 + (a'_{23} - a'_{N3})\, x_3 + \ldots = y'_2 - y'_N \\ \cdot \\ \cdot \\ (a'_{N-1,2} - a_{N2})\, x_2 + (a'_{N-1,3} - a'_{N3})\, x_3 + \ldots = y'_{N-1} - y'_N \end{array}\right\} \quad [2.2]$$

which is in the same form as equation [2.1] but now there is one less equation (the Nth having been reduced to a trivial $0 = 0$) and one less unknown (x_1 now being eliminated from all the $N - 1$ new equations).

Having gone from a system of N equations to a new system of $N - 1$ equations, we can treat [2.2] as we did [2.1] and reduce it to a system of $N - 2$ equations by, say, elimination of x_2. The process can be repeated until one unknown only, say x_N, is left. Once the value for x_N is known it can be substituted into one of the immediately preceding pair of equations to get x_{N-1}; the values for x_N and x_{N-1} are then put into one of the preceding set of three equations (involving x_{N-2}, x_{N-1}, x_N) and x_{N-2} is decided; proceeding in like manner, x_{N-3}, x_{N-4}, ..., and on back to x_1 are obtained in turn in a very straightforward way. This procedure can in practice be usefully applied in a computer to "invert" a system of linear equations, and it works well, provided precautions are taken to prevent division by zero or a very small number. Substitution of the solution for $x_1, x_2, ..., x_N$ into the left side of the set of equations [2.1] then gives numbers which are satisfactorily close to the quantities on the right side. In this sense the inversion of large linear systems presents no difficulties on a computer. But what of the obverse question — suppose a given set of values for $x_1, x_2, ..., x_N$ are combined with an array of coefficients $a_{11}, a_{12}, ..., a_{1N}; a_{21}, a_{22}, ..., a_{2N};$ . . . etc., and a set of values for $y_1, y_2, ..., y_N$ derived; when the inversion procedure is applied to the system of equations, how different are the solutions to the original values for x_1, x_2, etc? The answer is often — very far indeed!

This apparent contradiction arises because there are small errors present in almost every numerical calculation. Consider as an extreme case for illustration the simple pair of equations:

$$x + y = 2$$

$$2x + 2.000001y = 4.000001$$

The solution is $x = 1$, $y = 1$, and as long as the seven figure accuracy reflects the real facts there is no problem. But if there is any uncertainty either in the coefficients or on the right-hand side, a very different situation comes about. Suppose the equations are written without the small fractional term on the right; then they become:

$$x + y = 2$$

$$2x + 2.000001y = 4$$

whence, multiplying the first by two and subtracting from the second:

$$0.000001y = 0$$

Therefore $x = 2$, $y = 0$.

Suppose on the other hand that the coefficient 2.000001 is replaced by 2:

$$x + y = 2$$

$$2x + 2y = 4.000001$$

This implies the contradiction $0 = 0.000001$. Finally if both small decimal fractions are omited:

$$x + y = 2$$

$$2x + 2y = 4$$

The second equation of this pair merely repeats the same information as the first and there are an infinity of solutions; any pair of values for x and y such that $x + y = 2$ satisfies the system of equations.

It now becomes apparent that the simple nature of the problem becomes quite illustory when the system of equations are not exact statements of absolute fact. When finite errors, however small, are admitted either in the coefficients or on the right side of the system of equations, it may eventuate that the "solution" is more influenced by these errors than by the informative part of the equations. It is worth noting that the determinant of coefficients in the simple problem just given is small (0.000001) but does not vanish. Small values for the determinant are symptomatic of unstable inversions where very small changes lead to large changes in the solution.

Under- and over-determined systems of equations

The simple pairs of equations discussed above show the characteristics of "underdetermination" which become very important when larger systems of linear equations are encountered. If there are N unknowns and less than N equations, the system is said to be *underdetermined* — there is no single unique solution. The equations:

$$ax + by + cz = d$$

$$\alpha x + \beta y + \gamma z = \delta$$

for example, are satisfied by several sets of solutions:

(1) x arbitrary, any y and z connected by:

$$(\alpha b - a\beta)\, y + (\alpha c - a\gamma)\, z = \alpha d - a\delta$$

(2) y arbitrary, any x and z connected by:

$$(\beta a - b\alpha)\, x + (\beta c - b\gamma)\, z = \beta d - b\delta$$

and so on.

When there are more than N equations in N unknowns they may still not be overdetermined, for several of the equations may say the same thing; but if the equations are all independent — which means that none of them can be produced by a linear combination of the others — then a system of more than N equations in N unknowns must be in fact overdetermined. When the equations are not independent the system may turn out to be overdetermined, determinate or underdetermined, depending on how many redundant equations are present.

It is fairly simple to examine small systems of equations to determine whether they are underdetermined, overdetermined or determinate, but when the system is large this is no longer trivial. Furthermore, and more importantly, the simple example discussed earlier shows that very small changes in the numerical value of some coefficients in even a very small system of equations can change the character of the system from determinate to underdetermined or indeterminate (contradictory), so in the practical situation where measurements and approximations are involved (even though perhaps the measurements are accurate and the approximations very good), it is far from a trivial problem to ascertain the determinacy or otherwise of a linear system of equations. Evidently also, the conclusion cannot in many cases be an absolute one since we have seen that the results may depend on the accuracy with which the equations can be specified and the accuracy of a measured quantity has a statistical (rather than an absolute) significance: when a measurement error of say 2% is specified, this normally means that an error of 5% of 6% will occasionally occur, and even a 10% error is not impossible — it is merely very unlikely but will crop up at the one-in-a-million level.

2.2. QUADRATURE, THE REDUCTION OF INTEGRAL EQUATIONS TO SYSTEMS OF LINEAR EQUATIONS

Most inversion problems involve integral transforms $\int_a^b K_i(x)\, f(x)\, dx$, but numerical calculation for arbitrary $f(x)$ must rely on the approximation of this integral by a sum. Such approximation of an integral is known as "quadrature" or "numerical quadrature" and a variety of quadrature formulae exist. The accuracy of a quadrature can always be improved by increasing the number of points, i.e. the degree of subdivision in the interval (a, b). If $K_i(x)$ can change rapidly a very fine subdivision may be needed, but since most physical kernels in indirect sensing problems are quite smooth, a very large number of points is rarely needed to compute $\int K_i(x)\, f(x)\, dx$, given $f(x)$, to much better accuracy than that to which the transform can ever be measured.

A quite useful but simple quadrature procedure involves dividing the interval (a, b) into shorter intervals by interposing N so-called quadrature points $x_1, x_2, ..., x_N$ within the interval, envisaging $f(x)$ to take the values $f_1, f_2, ..., f_N$ at those points and to behave linearly across the subintervals $x_j \rightarrow x_{j+1}$ (Fig. 2.1). If $f(x)$ takes a fixed value (such as zero) at $x = a$ and $x = b$, the quadrature points need not include $x = a$ and $x = b$; otherwise x_1 will often be a and x_N will be b. In any event, the integral is approximated by summing over all the subintervals the integral $\int_{x_j}^{x_{j+1}} (A_j + B_j x)\, K_i(x)\, dx$ where A_j and B_j are chosen to make $A_j + B_j x$ coincide with $f(x)$ at $x = x_j$ and x_{j+1}.

To accomplish this:

$$A_j + B_j x_j = f_j$$

$$A_j + B_j x_{j+1} = f_{j+1}$$

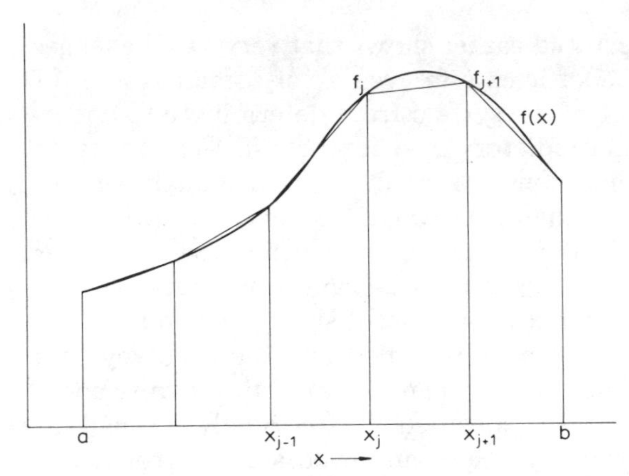

Fig. 2.1. A simple quadrature scheme.

So:

$$B_j = \frac{f_{j+1} - f_j}{x_{j+1} - x_j}, \qquad A_j = \frac{x_{j+1} f_j - x_j f_{j+1}}{x_{j+1} - x_j}$$

Thus the interval $x_j \to x_{j+1}$ makes the following contribution:

$$A_j \int_{x_j}^{x_{j+1}} K(x)\,\mathrm{d}x + B_j \int_{x_j}^{x_{j+1}} xK(x)\,\mathrm{d}x$$

which may be broken up in terms of factors of f_j and f_{j+1}:

$$\left[\frac{x_{j+1}}{x_{j+1} - x_j} \int_{x_j}^{x_{j+1}} K(x)\,\mathrm{d}x - \frac{1}{x_{j+1} - x_j} \int_{x_j}^{x_{j+1}} xK(x)\,\mathrm{d}x\right] \cdot f_j$$

$$+ \left[\frac{-x_j}{x_{j+1} - x_j} \int_{x_j}^{x_{j+1}} K(x)\,\mathrm{d}x + \frac{1}{x_{j+1} - x_j} \int_{x_j}^{x_{j+1}} xK(x)\,\mathrm{d}x\right] \cdot f_{j+1}$$

f_{j+1} will of course appear again from the interval $x_{j+1} \to x_{j+2}$ and f_j from the interval $x_{j-1} \to x_j$. Adding the contributions of all subintervals, one finally obtains the quadrature formula which can be written:

$$\int_a^b K(x)\, f(x)\,\mathrm{d}x \cong \sum_{j=1}^{N} w_j f_j \qquad [2.3]$$

where w_j represents:

$$\frac{-x_{j-1}}{x_j - x_{j-1}} \int_{x_{j-1}}^{x_j} K(x)\,\mathrm{d}x + \frac{x_{j+1}}{x_{j+1} - x_j} \int_{x_j}^{x_{j+1}} K(x)\,\mathrm{d}x + \frac{1}{x_j - x_{j-1}} \int_{x_{j-1}}^{x_j} xK(x)\,\mathrm{d}x$$

$$-\frac{1}{x_{j+1} - x_j} \int_{x_j}^{x_{j+1}} xK(x)\,\mathrm{d}x$$

The integrals $\int K(x)\,\mathrm{d}x$ and $\int xK(x)\,\mathrm{d}x$ for each subinterval may be evaluated numerically or analytically. The accuracy of the quadrature depends only on how good the linear approximation to $f(x)$ is across each interval.

The procedure outlined above is generally adequate for most numerical computations in remote sensing. More elaborate procedures exist (see, for example, Lanczos, 1956) but are not relevant to the present discussion since accuracy of *computing* $\int K(x) f(x)\,\mathrm{d}x$ is not usually a problem.

One can therefore by selecting a suitable quadrature formula reduce the integral equation:

$$g(y) = \int K(y, x) f(x)\,\mathrm{d}x \qquad [2.4]$$

to the linear approximation:

$$\underset{(M \times 1)}{g} = \underset{(M \times N)}{\mathbf{A}} \underset{(N \times 1)}{f} \qquad [2.5]$$

where f contains the values of $f(x)$ at a selected representative set of tabular points $x_1, x_2, \ldots, x_N$, $\boldsymbol{g}$ contains the M given values of $\boldsymbol{g}$, $g(y_1)$, $g(y_2)$, ..., $g(y_M)$, while a_{ij} is the jth quadrature coefficient for the integral containing the ith kernel, i.e.:

$$\int K(y_i, x) f(x)\,\mathrm{d}x \cong \sum_j a_{ij} f(x_j) \qquad [2.6]$$

In this way the integral equation is reduced to something qualitatively very similar to the simple puzzle quoted at the beginning of the chapter. But there are important quantitative differences — the numbers M and N are often large and the measured nature of $\boldsymbol{g}$ and the finite accuracy of quadrature formulae alone ensure that there will always be a small error component in the linear equation [2.4]; when we set out to invert that equation we invert not $\boldsymbol{g}$ but $\boldsymbol{g} + \epsilon$ where ϵ is an error term; we may have an estimate for the magnitude of ϵ but we do not know even the sign of the individual elements. The crucial question is what the effect of this uncertainty is likely to be on the solution. We have already seen that even in a very simple 2×2 system (2 equations, 2 unknowns), small numerical changes can dramatically alter the solution. It is obviously essential to examine this and related questions more closely and in a way which permits large (20×20, 50×40 or larger if necessary) systems to be handled.

BIBLIOGRAPHY

Lanczos, C., 1956. *Applied Analysis.* Prentice-Hall, Englewood Cliffs, N.J., 539 pp.

Ostrowsky, A.M., 1966. *Solutions of Equations and Systems of Equations.* Academic Press, New York, N.Y., 338 pp.

Wilkinson, J.H., 1963. *Rounding Errors in Algebraic Processes* (Notes on Applied Sciences No. 32, H.M. Stationery Office, London). Prentice-Hall, Englewood Cliffs, N.J., 161 pp.

CHAPTER 3

Theory of large linear systems

Before discussing further the questions raised in the previous chapter, it is desirable to introduce the matrix-vector notation which leads to a much more compact and powerful algebra — and also a much more readable presentation — than the cumbersome subscript notation used earlier. Matrix-vector theory carries over into complex numbers, but in the present context only real elements will be involved and consideration will be restricted to real arrays.

3.1. MATRIX-VECTOR ALGEBRA

Vectors

An ordered linear array of numbers is a vector. The numbers may be arranged in a horizontal or in a vertical direction and it is therefore usual to distinguish between "row vectors" and "column vectors". In all subsequent sections we will identify "vectors" with column vectors. Row vectors when they occasionally arise will be written as the "transpose" of column vectors. Transposition denotes the interchange of rows and columns — vectors will be represented by bold face, lower case, letters. Thus the (column) vector $\boldsymbol{v}$ may represent in concise notation the array:

$$\begin{bmatrix} v_1 \\ v_2 \\ v_3 \end{bmatrix}$$

The transpose $\boldsymbol{v}^*$ is $[v_1 \quad v_2 \quad v_3]$, a row vector

Matrices

Just as vectors are one-dimensioned arrays (single columns or single rows) involving a single subscript to identify the elements, matrices are two-dimensional arrays, involving double subscripts.

These arrays are envisaged to be arranged as a rectangular ordering in rows and columns. If there are m rows and n columns, the matrix is an $(m \times n)$ matrix. In some situations, a matrix $\mathbf{A}$ will be defined by using the notation

$||a_{ij}||$ to denote a matrix with general element a_{ij}, i.e. the ith row, jth column position in the matrix contains a_{ij}.

The transpose of **A** is then obviously $\mathbf{A}^*$[1] where:

$$\mathbf{A}^* = ||a_{ji}||$$

As another example of this notation, $||\int f_i(x)\, u_j(x)\, dx||$ is a matrix with the scalar given by the integral in the (i, j) position.

A single boldface capital will be the symbol mostly used for matrices in the rest of this book. For example, the symbol **A** may represent the matrix:

$$\begin{bmatrix} a_{11} & a_{12} & a_{13} \\ a_{21} & a_{22} & a_{23} \\ a_{31} & a_{32} & a_{33} \end{bmatrix}$$

However, the dimensions of the matrix (number of rows and columns) is not specified by the symbol **A**. If this is relevant, the number of rows and columns can be specified, say by writing $m \times n$ below the symbol to indicate that the matrix contains m rows and n columns.

Referring back to equation [2.1] we note that it involves an $(m \times n)$ matrix **A**, a vector $\boldsymbol{x}$ and a vector $\boldsymbol{y}$. Restricting consideration for the moment to the three-dimension situation ($n = 3$), one has:

$$x = \begin{bmatrix} x_1 \\ x_2 \\ x_3 \end{bmatrix} \qquad y = \begin{bmatrix} y_1 \\ y_2 \\ y_3 \end{bmatrix} \qquad \mathbf{A} = \begin{bmatrix} a_{11} & a_{12} & a_{13} \\ a_{21} & a_{22} & a_{23} \\ a_{31} & a_{32} & a_{33} \end{bmatrix}$$

Equation [2.1] written out in full becomes:

$$\left.\begin{aligned} a_{11}x_1 + a_{12}x_2 + a_{13}x_3 &= y_1 \\ a_{21}x_1 + a_{22}x_2 + a_{23}x_3 &= y_2 \\ a_{31}x_1 + a_{32}x_2 + a_{33}x_3 &= y_3 \end{aligned}\right\}$$

This can be written:

$$\mathbf{A}x = y$$

if we agree to define the product $\mathbf{A}x$ as a vector:

$$\begin{bmatrix} a_{11}x_1 & + & a_{12}x_2 & + & a_{13}x_3 \\ a_{21}x_1 & + & a_{22}x_2 & + & a_{23}x_3 \\ a_{31}x_1 & + & a_{32}x_2 & + & a_{33}x_3 \end{bmatrix}$$

[1] In common convention $\mathbf{A}^*$ denotes the Hermitian transpose wherein $\mathbf{A}^*$ contains in its (i, j) position not the (j, i) element of **A**, but its complex conjugate. With the real matrices to which we have restricted ourselves $\mathbf{A}^*$ is also the direct transpose.

or:

$$\begin{bmatrix} \sum_k a_{1k}x_k \\ \sum_k a_{2k}x_k \\ \sum_k a_{3k}x_k \end{bmatrix}$$

This then is the definition of premultiplication of a vector by a matrix. It is called premultiplication because the matrix appears first in the product. It may be noted that **A*x*** is *not* equal to ***x*A**: indeed, when ***x*** is a column vector ***x*A** has no meaning.

To form **A*x*** it is not necessary that **A** be square (i.e. that it have equal number of rows and columns). The definition of the matrix-vector product allows the premultiplication of a matrix with m rows and n columns (or, for brevity, an $m \times n$ matrix) by a vector with n elements (which of course can be described also as an $n \times 1$ matrix). Only the range of summation is involved — the product is:

$$\underset{(m \times n)}{\mathbf{A}} \; \underset{(n \times 1)}{\boldsymbol{x}} = \begin{bmatrix} \sum_{k=1}^{n} a_{1k}x_k \\ \sum_{k=1}^{n} a_{2k}x_k \\ \vdots \\ \sum_{k=1} a_{mk}x_k \end{bmatrix} \qquad [3.1]$$

Note that an $m \times n$ matrix can premultiply only an n-dimensioned column vector.

The product of an $m \times n$ matrix and an $n \times 1$ matrix (i.e., a vector) gives an $m \times 1$ matrix (also a vector but of dimension m). To form the first row or element of **A*x***, the first row of **A** and the column ***x*** are multiplied, element by element, with each other and summed. The second element (row) in the product is given by using the second row of **A**, and so on.

The product can also be described using the dot-product notation (borrowed from geometric vector analysis). The dot-product or inner product of two one-dimensional arrays of equal length $(u_1, u_2, \ldots u_n)$ and $(w_1, w_2, \ldots w_n)$ is the scalar quantity $\Sigma_1^n \, u_k w_k$.

In this notation, the mth row of **A*x*** is given by the dot-product of the

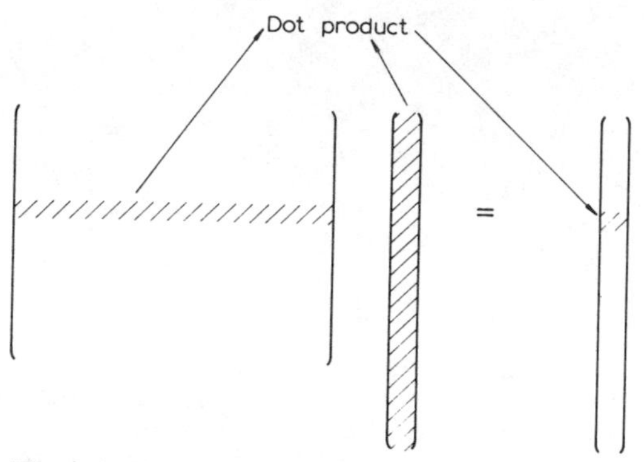

Fig. 3.1. Matrix-vector product.

mth *row* of $\mathbf{A}$ with the *column* array $\boldsymbol{x}$ (see Fig. 3.1). Both have dimension n.

3.2. MATRIX PRODUCTS

The product of an $m \times n$ matrix $\mathbf{B}$ with an n-dimensional vector $\boldsymbol{x}$ gives an m-dimensional vector $\mathbf{B}\boldsymbol{x}$. Suppose this in turn is premultiplied by an $(l \times m)$ matrix $\mathbf{A}$; l can be arbitrary, but the second dimension (number of columns in $\mathbf{A}$) evidently *must* be m. The vector thus formed is $\mathbf{A}(\mathbf{B}\boldsymbol{x})$.

The expression within parentheses is a vector and no new definitions are involved up to this point, but the parentheses can be dropped only if we suitably define a matrix product $\mathbf{AB}$. The definition must be consistent with the definition of a matrix-vector product. This consistency is easily achieved. Let $\mathbf{P}$ represent the matrix product; since it must give a l-dimensional vector when it premultiplies an n-dimension vector, this product must be an $l \times n$ matrix. Reverting for the moment to a subscript notation, the components of the final vector $\boldsymbol{v}$ are given by:

$$v_i = \sum_{j=1}^{n} p_{ij} x_j$$

and also by:

$$v_i = \sum_{j=1}^{m} a_{ij} \sum_{k=1}^{n} b_{jk} x_k$$

Evidently, consistency is achieved if the (i, k) element in the product is:

$$p_{ik} = \sum_{j=1}^{m} a_{ij} b_{jk} \qquad [3.2a]$$

or:

$$\mathbf{AB} = \| \sum_j a_{ij} b_{jk} \| \qquad [3.2b]$$

This is a fundamental, and very important relation, defining the matrix product. It is evident that the product of an $l \times m$ matrix and an $m' \times n$ matrix is an array of dot products of rows of the first matrix and columns of the second, and exists only when m and m' are equal. It also follows that the existence of **AB** does not even imply the existence of **BA**, much less that **AB** and **BA** (when it exists) are equal (Fig. 3.2). Hence matrix products are associative, but not commutative, i.e.:

$$\mathbf{A}(\mathbf{BC}) = (\mathbf{AB})\,\mathbf{C} \qquad [3.3]$$

but:

$$\mathbf{AB} \neq \mathbf{BA} \qquad [3.4]$$

The identity matrix **I** is defined by the relationship:

$$\mathbf{IA} = \mathbf{A}$$

for arbitrary **A**. It contains unity in all diagonal elements while all off-diagonal elements are zero. It is easy to show that **AI** must also equal **A**.

Outer and inner products of vectors

A vector $\boldsymbol{v}$ is also an $(n \times 1)$ matrix and matrix products can be formed by multiplication of vectors. These are known as inner or dot products when

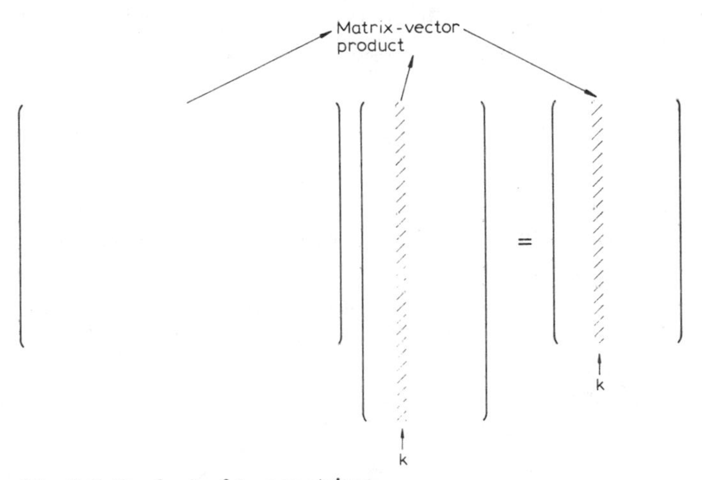

Fig. 3.2. Product of two matrices.

the premultiplier is a row vector $\boldsymbol{u}^*$ and the postmultiplier a column vector $\boldsymbol{v}$. The result of this — multiplication of a $(1 \times n)$ matrix by an $(n \times 1)$ matrix — gives a (1×1) matrix, i.e., a scalar $\boldsymbol{u}^* \boldsymbol{v}$.

If on the other hand, a column matrix $\boldsymbol{v}$, an $(n \times 1)$ matrix, premultiplies a row vector $\boldsymbol{u}^*$, a $(1 \times n)$ matrix, the result — the "outer product" — is an $n \times n$ matrix $\boldsymbol{vu}^*$. It contains in its (i, j) element the product of the ith element of $\boldsymbol{v}$ with the jth element of $\boldsymbol{u}$.

Thus for the three-dimensional case, we have:

$$\boldsymbol{u}^*\boldsymbol{v} = u_1v_1 + u_2v_2 + u_3v_3 = \boldsymbol{v}^*\boldsymbol{u}; \qquad \boldsymbol{uv}^* = \begin{bmatrix} u_1v_1 & u_1v_2 & u_1v_3 \\ u_2v_1 & u_2v_2 & u_2v_3 \\ u_3v_1 & u_3v_2 & u_3v_3 \end{bmatrix}$$

3.3 INVERSE OF A MATRIX

Premultiplication of a vector $\boldsymbol{x}$ by a matrix $\mathbf{A}$ gives a new vector, $\boldsymbol{y}$. The question immediately arises whether a matrix exists which premultiplying $\boldsymbol{y}$ gives $\boldsymbol{x}$. It is always possible to concoct such a matrix, but if the relationship $\mathbf{A}\boldsymbol{x} = \boldsymbol{y}$, $\mathbf{B}\boldsymbol{y} = \boldsymbol{x}$ holds for *all* $\boldsymbol{x}$ and $\boldsymbol{y}$, $\mathbf{B}$ is a matrix which inverts the effect of $\mathbf{A}$. It is then the inverse matrix of $\mathbf{A}$, denoted symbolically by $\mathbf{A}^{-1}$. Generally speaking, the inverse exists only when $\mathbf{A}$ is square and non-singular (the latter property implies that the determinant of the array $\mathbf{A}$, det $(\mathbf{A})$, is not zero). $\mathbf{AA}^{-1}$ evidently must equal $\mathbf{I}$.

The numerical computation of the inverse $\mathbf{A}^{-1}$ can be achieved on a computer by a number of methods, the most straightforward of which is merely an application of the elimination procedure applied earlier in the simple numerical examples in Chapter 2, of linear systems of equations. Consider, for example, equations [i], [ii] and [iii] (Chapter 2, section 2.1). Using matrix-vector notation, these may be written:

$$\overset{\mathbf{A}}{\begin{bmatrix} 1 & 1 & 1 \\ 2 & -1 & 3 \\ 3 & 2 & -1 \end{bmatrix}} \overset{\boldsymbol{x}}{\begin{bmatrix} x \\ y \\ z \end{bmatrix}} = \overset{\boldsymbol{y}}{\begin{bmatrix} 2 \\ 9 \\ -1 \end{bmatrix}}$$

In arriving at the solution, equation [i] was multiplied by 2 and subtracted from [ii]. These are now rows 1 and 2 of the matrix equation; the multiplication of row 1 by 2, followed by subtraction from row 2 to give a new second row is equivalent to premultiplication by a simple matrix $\mathbf{R}_1$:

$$\mathbf{R}_1 = \begin{bmatrix} 1 & 0 & 0 \\ -2 & 1 & 0 \\ 0 & 0 & 1 \end{bmatrix} \begin{matrix} \leftarrow \text{row 1 unchanged} \\ \leftarrow -2 \times \text{row 1} + 1 \times \text{ row 2 into row 2} \\ \leftarrow \text{row 3 unchanged} \end{matrix}$$

The next operation (row 3 replaced by row 3 minus 3 × row 1) is represented by:

$$\mathbf{R}_2 = \begin{bmatrix} 1 & 0 & 0 \\ 0 & 1 & 0 \\ -3 & 0 & 1 \end{bmatrix}$$

The total operation to this stage has been equivalent to premultiplication by $\mathbf{R}_2\mathbf{R}_1$. The next operation (the new row 3 multiplied by 3 and new row 2 subtracted, the result being put into row 3) is represented by:

$$\mathbf{R}_3 = \begin{bmatrix} 1 & 0 & 0 \\ 0 & 1 & 0 \\ 0 & -1 & +3 \end{bmatrix}$$

It is instructive at this point to calculate the elements of the matrix equation $\mathbf{R}_3\mathbf{R}_2\mathbf{R}_1\mathbf{A}\boldsymbol{v} = \mathbf{R}_3\mathbf{R}_2\mathbf{R}_1\boldsymbol{y}$. We have:

$$\mathbf{R}_3\mathbf{R}_2\mathbf{R}_1 = \begin{bmatrix} 1 & 0 & 0 \\ 0 & 1 & 0 \\ 0 & -1 & 3 \end{bmatrix} \begin{bmatrix} 1 & 0 & 0 \\ 0 & 1 & 0 \\ -3 & 0 & 1 \end{bmatrix} \begin{bmatrix} 1 & 0 & 0 \\ -2 & 1 & 0 \\ 0 & 0 & 1 \end{bmatrix} = \begin{bmatrix} 1 & 0 & 0 \\ 0 & 1 & 0 \\ 0 & -1 & 3 \end{bmatrix} \begin{bmatrix} 1 & 0 & 0 \\ -2 & 1 & 0 \\ -3 & 0 & 1 \end{bmatrix}$$

$$= \begin{bmatrix} 1 & 0 & 0 \\ -2 & 1 & 0 \\ -7 & -1 & 3 \end{bmatrix}$$

Thus:

$$\mathbf{R}_3\mathbf{R}_2\mathbf{R}_1\mathbf{A} = \begin{bmatrix} 1 & 0 & 0 \\ -2 & 1 & 0 \\ -7 & -1 & 3 \end{bmatrix} \begin{bmatrix} 1 & 1 & 1 \\ 2 & -1 & 3 \\ 3 & 2 & -1 \end{bmatrix} = \begin{bmatrix} 1 & 1 & 1 \\ 0 & -3 & 1 \\ 0 & 0 & -13 \end{bmatrix} = \mathbf{T}$$

This last matrix is an upper triangular matrix — all elements below the diagonal are zero, i.e. $t_{ij} = 0$ when $j < i$.

A triangular matrix is evidently simple to invert by the process of "back-substitution", which is a familiar operation in solving linear systems of equations. To solve an equation such as $\mathbf{T}\boldsymbol{v} = \boldsymbol{y}$, for example, one would proceed as follows. The last row of the equation is:

$$0 \cdot v_1 + 0 \cdot v_2 - 13 \cdot v_3 = y_3$$

whence $v_3 = -\frac{1}{13} y_3$.

This value is now substituted for v_3 into the preceding row:

$$0 \cdot v_1 - 3 \cdot v_2 + v_3 = y_2$$

to give:

$$v_2 = -\tfrac{1}{3} y_2 - \tfrac{1}{39} y_3$$

Substitution of the now known values for v_2 and v_3 convert the first row in its turn into an identity for v_1.

The process of back-substitution is readily described in a matrix notation. The various normalization and elimination steps which are taken can be represented as elementary matrix operations in each of which most of the off-diagonal elements are zero and most of the diagonal elements are unity. The main point is that an operation such as "multiply the kth row by α and subtract from β times the lth row, putting the result back as a new lth row" is a matrix operation, premultiplication by:

$$
\begin{array}{l}
\\ \\ \\ \\ \\
l\text{th} \rightarrow \\
\text{row} \\
\\ \\ \\
k\text{th} \rightarrow \\
\text{row} \\
\\ \\ \\ \\ \\
\end{array}
\left[\begin{array}{ccccccccccccccccc}
1 & 0 & & & & & & & & & & & & & & & \\
0 & \cdot & & & & & & & & & & & & & & & \\
 & & \cdot & & & & & & & & & & & & & & \\
 & & & 1 & & & & & & & & & & & & & \\
 & & & & 1 & & & & & & & & & & & & \\
0 & \cdot & \cdot & 0 & 0 & \beta & 0 & \cdot & \cdot & 0 & -\alpha & 0 & 0 & \cdot & \cdot & \cdot & 0 \\
 & & & \cdot & \cdot & 0 & 1 & 0 & \cdot & \cdot & 0 & & & & & & \\
 & & & & \cdot & \cdot & 0 & 1 & 0 & \cdot & 0 & & & & & & \\
 & & & & & \cdot & \cdot & 0 & 1 & 0 & 0 & & & & & & \\
 & & & & & & \cdot & \cdot & 0 & 1 & 0 & & & & & & \\
0 & & & & & & & \cdot & \cdot & 0 & 1 & 0 & \cdot & \cdot & \cdot & \cdot & \cdot \\
 & & & & & & & & & & & 1 & & & & & \\
 & & & & & & & & & & & & 1 & & & & \\
 & & & & & & & & & & & & & \cdot & & & \\
 & & & & & & & & & & & & & & \cdot & & \\
 & & & & & & & & & & & & & & & \cdot & \\
 & & & & & & & & & & \uparrow & & & & & & 1 \\
 & & & & & & & & & & k\text{th} & & & & & & \\
 & & & & & & & & & & \text{column} & & & & & &
\end{array}\right]
$$

Returning to the numerical values of **T**, the process of back-substitution in a matrix notation proceeds as follows: Divide the last row by $-\frac{1}{13}$ to obtain the last element z of the unknown vector $\boldsymbol{v}$; this is accomplished by any matrix containing $(0\ 0\ -\frac{1}{13})$ as its last row, for example:

$$
\begin{bmatrix} 1 & 0 & 0 \\ 0 & 1 & 0 \\ 0 & 0 & -\frac{1}{13} \end{bmatrix}
\begin{bmatrix} 1 & 1 & 1 \\ 0 & -3 & 1 \\ 0 & 0 & -13 \end{bmatrix}
\text{ is }
\begin{bmatrix} 1 & 1 & 1 \\ 0 & -3 & 1 \\ 0 & 0 & 1 \end{bmatrix}
$$

if now from the second row of the latter we subtract the last row, this is equivalent to premultiplication by a matrix with $(0, 1, -1)$ as its second row. To leave the other rows unchanged, we put unity in the diagonal elements and zero off diagonal:

$$\begin{bmatrix} 1 & 0 & 0 \\ 0 & 1 & -1 \\ 0 & 0 & 1 \end{bmatrix} \begin{bmatrix} 1 & 1 & 1 \\ 0 & -3 & 1 \\ 0 & 0 & 1 \end{bmatrix} \text{ is } \begin{bmatrix} 1 & 1 & 1 \\ 0 & -3 & 0 \\ 0 & 0 & 1 \end{bmatrix}$$

The −3 can be made 1 by a premultiplication:

$$\begin{bmatrix} 1 & 0 & 0 \\ 0 & -\frac{1}{3} & 0 \\ 0 & 0 & 1 \end{bmatrix} \begin{bmatrix} 1 & 1 & 1 \\ 0 & -3 & 0 \\ 0 & 0 & 1 \end{bmatrix} \text{ is } \begin{bmatrix} 1 & 1 & 1 \\ 0 & 1 & 0 \\ 0 & 0 & 1 \end{bmatrix}$$

Addition of the last two rows and subtraction of this sum from the first rows makes the first row (1 0 0):

$$\begin{bmatrix} 1 & -1 & -1 \\ 0 & 1 & 0 \\ 0 & 0 & 1 \end{bmatrix} \begin{bmatrix} 1 & 1 & 1 \\ 0 & 1 & 0 \\ 0 & 0 & 1 \end{bmatrix} \text{ is } \begin{bmatrix} 1 & 0 & 0 \\ 0 & 1 & 0 \\ 0 & 0 & 1 \end{bmatrix} = \mathbf{I}$$

The full sequence of operations which reduced the upper triangular matrix **T** to the identity matrix is described by:

$$\underset{\mathbf{R}_7}{\begin{bmatrix} 1 & -1 & -1 \\ 0 & 1 & 0 \\ 0 & 0 & 1 \end{bmatrix}} \underset{\mathbf{R}_6}{\begin{bmatrix} 1 & 0 & 0 \\ 0 & -\frac{1}{3} & 0 \\ 0 & 0 & 1 \end{bmatrix}} \underset{\mathbf{R}_5}{\begin{bmatrix} 1 & 0 & 0 \\ 0 & 1 & -1 \\ 0 & 0 & 1 \end{bmatrix}} \underset{\mathbf{R}_4}{\begin{bmatrix} 1 & 0 & 0 \\ 0 & 1 & 0 \\ 0 & 0 & -\frac{1}{13} \end{bmatrix}} \underset{\mathbf{T}}{\begin{bmatrix} 1 & 1 & 1 \\ 0 & -3 & 1 \\ 0 & 0 & -13 \end{bmatrix}} = \mathbf{I}$$

The last matrix on the right is $\mathbf{R}_3\mathbf{R}_2\mathbf{R}_1\mathbf{A}$, so we have:

$$(\mathbf{R}_7\mathbf{R}_6\mathbf{R}_5\mathbf{R}_4\mathbf{R}_3\mathbf{R}_2\mathbf{R}_1)\,\mathbf{A} = \mathbf{I}$$

Hence the inverse of **A** is:

$$\mathbf{A}^{-1} = \mathbf{R}_7\mathbf{R}_6\mathbf{R}_5\mathbf{R}_4\mathbf{R}_3\mathbf{R}_2\mathbf{R}_1$$

$\mathbf{R}_3\mathbf{R}_2\mathbf{R}_1$ has earlier been calculated explicitly as:

$$\begin{bmatrix} 1 & 0 & 0 \\ -2 & 1 & 0 \\ -7 & -1 & 3 \end{bmatrix}$$

$\mathbf{R}_7\mathbf{R}_6\mathbf{R}_5\mathbf{R}_4$ can be calculated fairly easily, being:

$$\begin{bmatrix} 1 & \frac{1}{3} & \frac{4}{39} \\ 0 & -\frac{1}{3} & -\frac{1}{39} \\ 0 & 0 & -\frac{1}{13} \end{bmatrix}$$

So $\mathbf{A}^{-1}$ is given by:

$$\mathbf{A}^{-1} = \begin{bmatrix} 1 & \frac{1}{3} & \frac{4}{39} \\ 0 & -\frac{1}{3} & -\frac{1}{39} \\ 0 & 0 & -\frac{1}{13} \end{bmatrix} \begin{bmatrix} 1 & 0 & 0 \\ -2 & 1 & 0 \\ -7 & -1 & 3 \end{bmatrix} = \frac{1}{39} \begin{bmatrix} -15 & 9 & 12 \\ 33 & -12 & -3 \\ 21 & 3 & -9 \end{bmatrix}$$

From the point of view simply of solving the equation $\mathbf{A}\boldsymbol{v} = \boldsymbol{y}$, the above procedure is simply a longwinded version of the well-known algebraic elimination procedure. We have accomplished in addition to "remember" all the simple operations used in elimination and to collect them into a single compact operator ($\mathbf{A}^{-1}$) which will allow $\mathbf{A}\boldsymbol{v} = \boldsymbol{y}$ to be solved for any other $\boldsymbol{y}$ without repeating the detailed steps.

To assist the interested reader in verifying that this procedure has succeeded, we show below the first row of $\mathbf{A}$, written as a column vector, and alongside it the three columns of our computed $\mathbf{A}^{-1}$ (omitting the common scalar factor 1/39):

$$\underset{r_1}{\begin{bmatrix} 1 \\ 1 \\ 1 \end{bmatrix}} \underset{c_1}{\begin{bmatrix} -15 \\ 33 \\ 21 \end{bmatrix}} \underset{c_2}{\begin{bmatrix} 17 \\ -14 \\ -3 \end{bmatrix}} \underset{c_3}{\begin{bmatrix} 12 \\ -3 \\ -9 \end{bmatrix}}$$

It is easily verified that the dot product $\boldsymbol{r}_1 \cdot \boldsymbol{c}_1$ is 39, while the other columns give dot products with $\boldsymbol{r}_1$ which vanish. The second row of $\mathbf{A}$ is:

$$\boldsymbol{r}_2 = \begin{bmatrix} 2 \\ -1 \\ 3 \end{bmatrix}$$

and we have $\boldsymbol{r}_2 \cdot \boldsymbol{c}_1 = 0, \boldsymbol{r}_2 \cdot \boldsymbol{c}_2 = 39, \boldsymbol{r}_2 \cdot \boldsymbol{c}_3 = 0$. The third row of $\mathbf{A}$ is:

$$\boldsymbol{r}_3 = \begin{bmatrix} 3 \\ 2 \\ -1 \end{bmatrix}$$

and gives $\boldsymbol{r}_3 \cdot \boldsymbol{c}_1 = 0, \boldsymbol{r}_3 \cdot \boldsymbol{c}_2 = 0, \boldsymbol{r}_3 \cdot \boldsymbol{c}_3 = 39$.

These row-column dot products are simply the elements of the matrix product (the m, n element of the matrix product being given by the dot product of the mth row of the premultiplying matrix with the nth column of the postmultiplying matrix). Hence:

$$\begin{bmatrix} 1 & 1 & 1 \\ 2 & -1 & 3 \\ 3 & 2 & 1 \end{bmatrix} \begin{bmatrix} -15 & 7 & 12 \\ 33 & -14 & -3 \\ 21 & -3 & -9 \end{bmatrix} = \begin{bmatrix} 39 & 0 & 0 \\ 0 & 39 & 0 \\ 0 & 0 & 39 \end{bmatrix} = 39\mathbf{I}$$

So the inverse has been correctly calculated.

The foregoing procedure can be applied directly for computer calculation of inverse matrices. When the matrices are large there is the possibility that some elements may become very small or even zero and it is desirable to avoid division by such small elements. This can be done quite simply if the largest off-diagonal element is reduced to zero at each stage. If the largest (the so-called "pivotal" element) is, say, a_{12}, the diagonal element below it in the same column is a_{22}; multiplication of the second row by $\mathbf{B} = a_{12}/a_{22}$ gives a row of numbers with a_{12} in the second column, subtraction of that row from the first puts a zero in the position formerly occupied by the largest, pivotal element a_{12}. Multiplication of row 2 by B and subtraction from row 1 (leaving the original row 2 unaltered) is equivalent to premultiplication of A by a matrix:

$$\mathbf{R}_1 = \begin{bmatrix} 1 & -\beta & 0 & 0 & . & . \\ 0 & 1 & 0 & . & . & . \\ 0 & 0 & 1 & . & . & . \\ . & . & . & 1 & . & . \\ . & . & . & . & . & . \end{bmatrix}$$

Thus the identities:

$$\left.\begin{array}{llll} a_{11}X_1 & + a_{12}X_2 & + a_{13}X_3 & + \ldots = Y_1 \\ a_{21}X_1 & + a_{22}X_2 & + a_{23}X_3 & + \ldots = Y_2 \\ . & . & . & . \\ . & . & . & . \end{array}\right\}$$

are replaced by the identities:

$$\left.\begin{array}{llll} (a_{11} - \beta a_{21})\, X_1 & + 0 \cdot X_2 + (a_{13} - \beta a_{23})\, X_3 & + \ldots & = Y_1 - \beta y_2 \\ a_{21}X_1 & + a_{22}X_2 + a_{23}X_3 & + \ldots & = Y_2 \\ . & . \quad . & & . \\ . & . \quad . & & . \end{array}\right\}$$

Or, in matrix notation:

$$\mathbf{A}x = y$$

is replaced by:

$$\mathbf{R}_1\mathbf{A}x = \mathbf{R}_1 y$$

The process of solution by elimination is complete when the left-hand matrix has been changed from the original **A** to the identity matrix **I.** If this is achieved after m steps, the result can be written:

$$\mathbf{R}_m\mathbf{R}_{m-1} \ldots \mathbf{R}_1\mathbf{A}x = Ix = x = \mathbf{R}_m\mathbf{R}_{m-1} \ldots \mathbf{R}_1 y$$

So by repeated eliminations not only has x been solved for but the inverse matrix $\mathbf{A}^{-1}$ is given by $(\mathbf{R}_m \mathbf{R}_{m-1} \ldots \mathbf{R}_1)$.

If the same A is involved in another problem, say:

$$\mathbf{A}u = v$$

Then if $\mathbf{A}^{-1}$, i.e. $(\mathbf{R}_m \mathbf{R}_{m-1} \ldots \mathbf{R}_1)$, has been retained, u is given by $\mathbf{A}^{-1}v$, i.e. a single matrix-vector multiplication suffices to obtain the solution vector.

3.4. TRANSPOSITION AND RULES FOR PRODUCT INVERSION

The transpose $\mathbf{A}^*$ of a matrix $\mathbf{A}$ is simply obtained by interchanging i, j elements and j, i elements in the matrix or (equivalently) by interchanging rows and columns. The diagonal elements are of course unaffected by tranposition. A *symmetric matrix* has the property $\mathbf{A}^* = \mathbf{A}$.

The transposition of the product $\mathbf{P}$ of two matrices $\mathbf{A}$ and $\mathbf{B}$ is readily written down if we recall the rule for forming a matrix product:

$$p_{ij} = \sum_k a_{ik} b_{kj}; \qquad p_{ji} = \sum a_{jk} b_{ki}$$

but b_{ki} is the i, k element in the transpose $\mathbf{B}^*$ while a_{jk} is the k, j element in $\mathbf{A}^*$. Hence:

$$p^*_{ij} = p_{ji} = \sum_k b^*_{ik} a^*_{kj}$$

Or:

$$\mathbf{P}^* = \mathbf{B}^* \mathbf{A}^*$$

Or:

$$(\mathbf{AB})^* = \mathbf{B}^* \mathbf{A}^* \qquad [3.5]$$

This characteristic reversal of order is found also in forming inverses, that is:

$$(\mathbf{AB})^{-1} = \mathbf{B}^{-1} \mathbf{A}^{-1} \qquad [3.6]$$

which can readily be seen if we note that the relationships:

$$(\mathbf{AB})^{-1} \mathbf{AB} = \mathbf{I}$$

$$\mathbf{AB}(\mathbf{AB})^{-1} = \mathbf{I}$$

plus the associative property of matrix products (which was made to hold by the definition of that product) lead immediately to equation [3.6] — if, for example, we *post*multiply the first relationship by $\mathbf{B}^{-1}$ to obtain:

$$(\mathbf{AB})^{-1} \mathbf{A} = \mathbf{B}^{-1}$$

and then *post*multiply both sides by $\mathbf{A}^{-1}$:

$$(\mathbf{AB})^{-1} = \mathbf{B}^{-1}\mathbf{A}^{-1}$$

In a similar way, we can readily prove that the inverse of **A** gives the identity matrix when premultiplied *or* postmultiplied by **A**. As derived earlier, the property of the inverse was:

$$\mathbf{A}^{-1}\mathbf{A} = \mathbf{I}$$

whence $(\mathbf{AA}^{-1})\,\mathbf{A} = \mathbf{A}$ or:

$$\mathbf{AA}^{-1} = \mathbf{I}$$

The symmetric product A^*A

This product will be encountered frequently. When **A** is asymmetric or even non-square, the product $\mathbf{A}^*\mathbf{A}$ is symmetric and square. The square nature follows from the fact that if **A** is an $(m \times n)$ matrix $\mathbf{A}^*$ is an $(n \times m)$ matrix, so $\mathbf{A}^*\mathbf{A}$ is an $(n \times n)$ matrix. The symmetry of $\mathbf{A}^*\mathbf{A}$ follows immediately from the transposition formula:

$$(\mathbf{A}^*\mathbf{A})^* = \mathbf{A}^*(\mathbf{A}^*)^*$$

but the double transposition expressed by $(\mathbf{A}^*)^*$ must restore the initial **A**, hence $\mathbf{A}^*\mathbf{A}$ is unaffected by transposition and therefore symmetric. ($\mathbf{AA}^*$ is an $m \times m$ square symmetric matrix, and so is not the same as $\mathbf{A}^*\mathbf{A}$.)

If **A** itself is square, the inverse of **A** can be calculated from the inverse of $\mathbf{A}^*\mathbf{A}$, for we have:

$$(\mathbf{A}^*\mathbf{A})^{-1} = \mathbf{A}^{-1}(\mathbf{A}^*)^{-1}$$

So:

$$\mathbf{A}^{-1} = (\mathbf{A}^*\mathbf{A})^{-1}\,\mathbf{A}^* \qquad [3.7]$$

— that is, a simple postmultiplication by $\mathbf{A}^*$ makes $(\mathbf{A}^*\mathbf{A})^{-1}$ into $\mathbf{A}^{-1}$. **A**, of course, may be asymmetric and this result enables any real matrix to be inverted by procedures which can be applied directly only to symmetric matrices.

Decomposition of a real symmetric matrix into a product of triangular matrices

Many of the matrices encountered in indirect sensing are real symmetric matrices. A special property of such matrices is their reducibility to a product of triangular matrices one lower triangular (i.e. all elements above the diagonal are zero) and one upper diagonal (all below-diagonal elements zero).

Note that the product of two *lower* diagonal matrices can only give another *lower* diagonal matrix (for $\Sigma_k\ t_{ik}t'_{kj} = \Sigma_1^i\ t_{ik}t'_{kj} = 0$ if $j > i$), so the product involves both kinds of triangular matrix. When we premultiply a real lower triangular matrix $\mathbf{T}_\mathrm{L}$ by the upper triangular matrix $\mathbf{T}_\mathrm{U} = \mathbf{T}_\mathrm{L}{}^*$ formed by transposition of $\mathbf{T}_\mathrm{L}$ we obtain the matrix (Fig. 3.3):

$$\mathbf{A} = \mathbf{T}_\mathrm{L}\mathbf{T}_\mathrm{L}^* = \mathbf{T}_\mathrm{L}\mathbf{T}_\mathrm{U} = \mathbf{T}_\mathrm{U}^*\mathbf{T}_\mathrm{U}$$

which evidently is symmetric.

To show that symmetry of $\mathbf{A}$ is sufficient to enable it to be factored into products of triangular matrices, it is necessary only to note that the elements of $\mathbf{A}$ in the above equation are given by the dot products of the rows of $\mathbf{T}_\mathrm{L}$ — the i, j element of $\mathbf{A}$, being the dot product of the ith row of $\mathbf{T}_\mathrm{L}$ and the jth column of $\mathbf{T}_\mathrm{L}^*$, i.e. the dot product of the ith and jth rows of $\mathbf{T}_\mathrm{L}$. The (1, 1) element in $\mathbf{A}$ is simply t_{11}^2 and providing a_{11} is positive (one can always take out a factor of −1 if necessary), t_{11} is simply $\sqrt{a_{11}}$. The other elements of the first row of A are the dot products of the first row r_1 of $\mathrm{T_L}$ (which is $t_{11}, 0, 0, \ldots 0$) with the other rows of $\mathbf{T}_\mathrm{L}$, taken in turn, a_{1k} being $\boldsymbol{r}_1 \cdot \boldsymbol{r}_k = t_{11}t_{k1}$; thus, if we choose for t_{k1} the value $a_{1k}a_{11}^{-1/2}$, the relation $\mathbf{A} = \mathbf{T}_\mathrm{L}\mathbf{T}_\mathrm{L}^*$ is satisfied so far as the first row of $\mathbf{A}$ is concerned and the first column of T has been computed:

$$t_{11} = a_{11}^{1/2}; \qquad t_{k1} = a_{1k}a_{11}^{-1/2} \qquad (k \neq 1)$$

The second row of $\mathbf{A}$ involves dot products of the second row of t with the other rows; a_{21} of course is already decided since it equals a_{12}. The

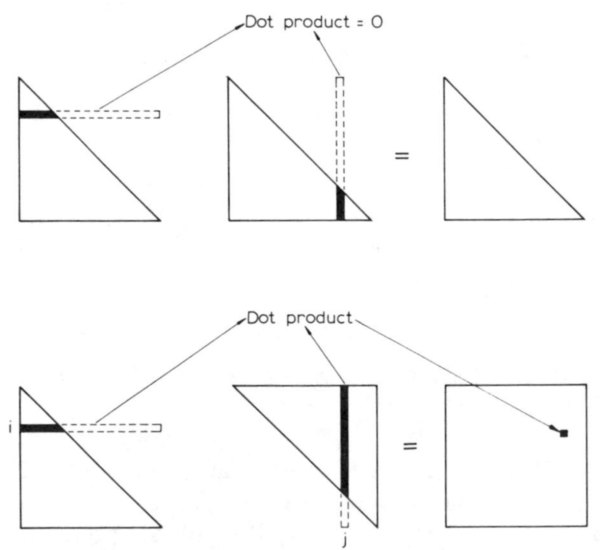

Fig. 3.3. Products of triangular matrices.

diagonal element a_{22} is $t_{21}^2 + t_{22}^2$; t_{21} having already been decided (the first column of t_{21} being computed at the previous stage), it follows that t_{22} must be $(a_{22} - t_{21}^2)^{1/2}$ — again a negative number may have to be made positive by factoring out -1 from the entire row of **A**. The remaining elements of the second row of **A** give the rest of the second column of $\mathbf{T}_L$, since:

$$a_{2k} = r_2 \cdot r_k = t_{21}t_{k1} + t_{22}t_{k2}$$

so that:

$$t_{k2} = (a_{2k} - t_{21}t_{k1})(a_{22} - t_{21}^2)^{-1/2} \qquad (k \neq 2)$$

Proceeding row by row in **A**, one computes first a diagonal element of t and then the rest of the non zero elements in sequence. If no negative numbers had to be altered by factoring out -1, the decomposition of **A** is complete and we have $\mathbf{A} = \mathbf{T}_L\mathbf{T}_L^*$. Otherwise what we have obtained is a decomposition of the matrix **DA** where **D** is a diagonal matrix with -1 in the positions where a (-1) was factored out of **A**, and with $+1$ in the other diagonal elements.

This provides the basis for an inversion method for symmetric real matrices, known for obvious reasons as the "square root" method. One first decomposes **DA** as has just been done, then since:

$$\mathbf{A} = \mathbf{D}\mathbf{T}_L\mathbf{T}_L^*$$

we have:

$$\mathbf{A}^{-1} = (\mathbf{T}_L^{-1})^* \ (\mathbf{T}_L)^{-1} \ \mathbf{D}$$

($\mathbf{D}^{-1}$ equals **D** because $\mathbf{D}^2 = \mathbf{I}$). To complete the inversion, all that is needed is the inversion of one of the triangular matrices, which has been described earlier in this chapter.

The square root method can be applied immediately only to real symmetric matrices, but the simple and rather surprising result given above (equation [3.8]) enables it to be used also to obtain the inverse of asymmetric real matrices, because if $(\mathbf{A}^*\mathbf{A})^{-1}$ is postmultiplied by $\mathbf{A}^*$, the result is $\mathbf{A}^{-1}$.

The main purpose of this chapter has been the recounting of results, definitions and properties which will be utilized in later chapters. It represents a selective extract from a much more extensive theory. The interested reader is referred to more exhaustive texts on matrix algebra for further discussions.

Most of the theory of matrices is quite old, tracing back to the great mathematicians of the 19th and 18th centuries. It existed, however, as a rather abstract body of symbolic theory until present-day computers made it possible to actually carry out numerically the operations of matrix multiplication, inversion and so on, which were previously well understood but never done explicitly (except for very small matrices) because the computational effort needed even for inversion of a (5×5) matrix or even multiplication of

(10×10) matrices is very great indeed. It is exceedingly instructive to attempt the inversion of a (5×5) matrix in numerical form:

BIBLIOGRAPHY

Aitken, A.C., 1956. *Determinants and Matrices.* Interscience, New York, N.Y., 144 pp.

Fedeeva, V.N., 1959. *Computational Methods of Linear Algebra.* Dover, New York, N.Y., 252 pp.

Forsyth, G.E., 1953. Solving linear algebraic equations can be interesting. *Bull. Am. Math. Soc.*, 59: 299–329.

Fox, L., 1964. *An Introduction to Numerical Linear Algebra.* Oxford University Press, Oxford 295 pp.

Hamming, R.W., 1973. *Numerical Methods for Scientists and Engineers.* McGraw-Hill, New York, N.Y., 721 pp.

Householder, A.S., 1964. *The theory of Matrices in Numerical Analysis.* Blaisdell, New York, N.Y., 257 pp.

Lanczos, C., 1956. *Applied Analysis.* Prentice-Hall, Englewood Cliffs, N.J., 539 pp.

Marcus, M., 1960. *Basic Theorems in Matrix Theory* (Applied Mathematics Series, N.B.S.). U.S. Government Printing Office, Washington, D.C., 59 pp.

CHAPTER 4

Physical and geometric aspects of vectors and matrices

It is easy to define matrix and vector products as we have just done but to the reader this will not often be very satisfying — why, he may ask, this particular definition, and he may envisage other operations involving arrays of numbers which equally well could be defined as matrix products. For example, why not define the product of two matrices so that a particular element in the product matrix is the product of the corresponding elements in the matrices being multiplied?

To explain or, at least, justify the definitions it is necessary to apply matrices and vectors to some problem, even if it is a contrived one. One could invent an algebra for just about any definition of multiplication and the definitions given earlier are just one possible choice. Their reasonableness, relevance and consistency we will attempt to demonstrate by example.

4.1. GEOMETRIC VECTORS

The very name "vector" has a geometric connotation and the parallel between a geometric vector and a set of three angles or three coordinates is obvious. We may specify position with respect to some rectangular axes by writing, for example,

$$x = 3\,; \qquad y = -2\,; \qquad z = 10$$

to specify the position of a point in space, or to categorize the radius vector from the origin of the coordinate system (i.e. the point where $x = 0, y = 0, z = 0$). This can be viewed as a single algebraic vector with elements 3, −2, 10. If we denote that vector by $\boldsymbol{r}$, the distance of the point from the origin, or the length of the radius vector to that point is evidently given by:

$$r^2 = \boldsymbol{r}^*\boldsymbol{r} = \boldsymbol{r} \cdot \boldsymbol{r}$$

The distance between point (3, −2, 10) and point (1, 4, 5) is given by $\sqrt{(3-1)^2 + (4-[-2])^2 + (10-5)^2}$. If we use only symbols we may write a distance as $\sqrt{(x-\xi)^2 + (y-\eta)^2 + (z-\zeta)^2}$, where the distance is measured between two points categorized by coordinates (x, y, z) and (ξ, η, ζ), respectively.

We can, of course, change our coordinate system by changing the origin and/or by rotating the axes. If we only change the origin, say to $(x_0, y_0,$

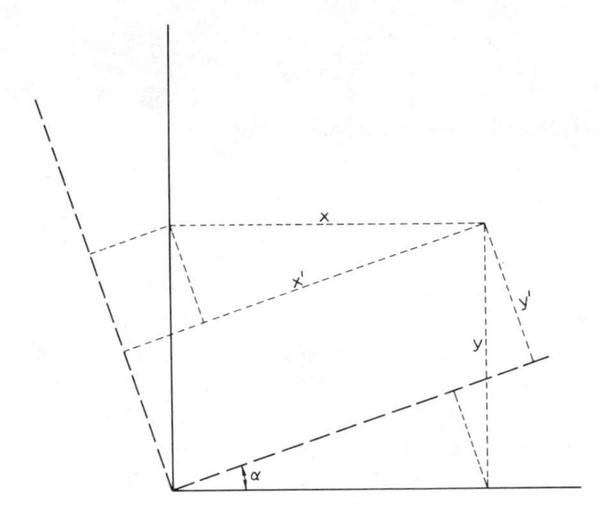

Fig. 4.1. Rotation of axes.

z_0) leaving the direction of the axes unchanged, then with respect to this new origin the coordinates become $(x - x_0, y - y_0, z - z_0)$; the new x-values, do not involve the old y- or z-values, only the old x-values. But if we allow the reference axes to rotate, the situation is less simple. To take the simplest case of rotation first, suppose in Fig. 4.1 the solid axes are the original axes but we wish to change over to a new system of axes which are directed along the broken lines at an angle α to the old axes. The z-axis we will suppose to be unchanged.

If x, y are the old values, the construction shown in the figure enables us to write down immediately the new values which are, respectively:

$$\left.\begin{aligned} x' &= x \cos\alpha + y \sin\alpha \\ y' &= -x \sin\alpha + y \cos\alpha \end{aligned}\right\}$$

Recalling our matrix-vector definitions, we see that the changed coordinates can be written as a matrix-vector equation:

$$\begin{bmatrix} x' \\ y' \end{bmatrix} = \begin{bmatrix} \cos\alpha & \sin\alpha \\ -\sin\alpha & \cos\alpha \end{bmatrix} \begin{bmatrix} x \\ y \end{bmatrix}$$

or, if we wish to include the z-coordinate:

$$\begin{bmatrix} x' \\ y' \\ z' \end{bmatrix} = \begin{bmatrix} \cos & \sin & 0 \\ -\sin & \cos & 0 \\ 0 & 0 & 1 \end{bmatrix} \begin{bmatrix} x \\ y \\ z \end{bmatrix}$$

Thus we see that the process of rotation can be represented in a simple way by a matrix which acts on the old coordinates, collected into a single vector $\boldsymbol{r}$, to give the new coordinates, contained in the vector $\boldsymbol{r}'$ (every matrix cannot be identified with some kind of rotation, however).

In physics we find many directed quantities, or vectors. Electric field $\boldsymbol{E}$, for example has a direction and can be described by giving the components E_x, E_y, E_z along arbitrary orthogonal x-, y- and z-axes. Other quantities, such as the dipole moment $\boldsymbol{p}$ induced in a body by the electric field are also directed quantities. If the body is "isotropic" the dipole moment is always parallel to $\boldsymbol{E}$ and so can be written as $\boldsymbol{p} = a\boldsymbol{E}$, where a is some constant of proportionality and a single number. But in some cases the body may not be as readily polarized in one direction as in another, in which case we need at least three proportionality constants a_x, a_y, a_z to describe the relationship between $\boldsymbol{p}$ and $\boldsymbol{E}$; if:

$$p_x = a_x E_x \,; \qquad p_y = a_y E_y \,; \qquad p_z = a_z E_z$$

which can be written in a matrix-vector form:

$$\boldsymbol{p} = \begin{bmatrix} p_x \\ p_y \\ p_z \end{bmatrix} = \begin{bmatrix} a_x & 0 & 0 \\ 0 & a_y & 0 \\ 0 & 0 & a_z \end{bmatrix} \begin{bmatrix} E_x \\ E_y \\ E_z \end{bmatrix} = \mathbf{A}\boldsymbol{E}$$

which still seems a somewhat forced and artificial notation. But suppose that E_x can contribute to the dipole moment in the y- and/or z-directions, E_y to the moment in the z- and x-directions and E_z to the x- and y-moments: in that case the simplest possible relationship is:

$$\boldsymbol{p} = \begin{bmatrix} p_x \\ p_y \\ p_z \end{bmatrix} = \begin{bmatrix} a_{xx} & a_{xy} & a_{xz} \\ a_{yx} & a_{yy} & a_{yz} \\ a_{zx} & a_{zy} & a_{zz} \end{bmatrix} \begin{bmatrix} E_x \\ E_y \\ E_z \end{bmatrix}$$

— we are then, in fact, forced to use a matrix to describe the relationship. But what of the simpler relationship described by the equations $p_x = a_x E_x$; $p_y = a_y E_y$; $p_z = a_z E_z$? So long as the x-, y-, z-axes coincide with some *special* directions in the body itself, such as axes with respect to which certain symmetries exist, the idea of separate proportionality constants is acceptable. Let us suppose that x, y, z happen to be so chosen, but now let us transform to another coordinate system, using first the simple rotation in the x, y-plane discussed earlier, which gave the matrix:

$$\begin{bmatrix} \cos\alpha & \sin\alpha & 0 \\ -\sin\alpha & \cos\alpha & 0 \\ 0 & 0 & 1 \end{bmatrix}$$

or, equivalently the linear equations:

$$x' = \cos\alpha \cdot x + \sin\alpha \cdot y + 0 \cdot z$$

$$y' = -\sin\alpha \cdot x + \cos\alpha \cdot y + 0 \cdot z$$

$$z' = 0 \quad + 0 \quad + z$$

to describe the effects of the rotation on the coordinates x, y, z. It is easily seen that the components of any other directed quantity must be recalculated for the new axes in exactly the same way, i.e. the new components for the electric field are:

$$\left.\begin{array}{l} E'_x = \cos\alpha \cdot E_x + \sin\alpha \cdot E_y + 0 \\ E'_y = -\sin\alpha \cdot E_x + \cos\alpha \cdot E_y + 0 \\ E'_z = 0 \quad + 0 \quad + E_z \end{array}\right\}$$

the electric field is still the same field, physically; we are merely measuring the components in different directions to those which we formerly employed. For the dipole moment we have:

$$p'_x = \cos\alpha \cdot p_x + \sin\alpha \cdot p_y = a_x \cos\alpha \cdot E_x + a_y \sin\alpha \cdot E_y$$

$$p'_y = -\sin\alpha \cdot p_x + \cos\alpha \cdot p_y = -a_x \sin\alpha \cdot E_x + a_y \cos\alpha \cdot E_y$$

$$p'_z = \quad p_z = \quad a_z E_z$$

but this relates $\boldsymbol{p}'$ (in the *new* coordinates) to $\boldsymbol{E}$ (in the *old*). We have not explicitly derived a way to express the old E_x, E_y, E_z in terms of the new, but it obviously corresponds to a rotation of $-\alpha$ in the x, y plane, and can be written down:

$$\left.\begin{array}{l} E_x = \cos\alpha \cdot E'_x - \sin\alpha E'_y \\ E_y = \sin\alpha \cdot E'_x + \cos\alpha E' \\ E_z = 0 \quad + 0 \quad + E'_z \end{array}\right\}$$

That this is correct can be verified by writing in the values $\cos\alpha\, E_x + \sin\alpha\, E_y$, etc., for E'_x, E'_y, E'_z:

$$E_x = \cos\alpha(\cos\alpha E_x + \sin\alpha E_y) - \sin\alpha(-\sin\alpha E_x + \cos\alpha E_y)$$

$$= (\cos^2\alpha + \sin^2\alpha)\, E_x + (\cos\alpha \sin\alpha - \cos\alpha \sin\alpha)\, E_y$$

$$= E_x\,,$$

and so on; or it can be verified much more succinctly by establishing that the matrices:

$$\begin{bmatrix} \cos\alpha & \sin\alpha & 0 \\ -\sin\alpha & \cos\alpha & 0 \\ 0 & 0 & 1 \end{bmatrix} \begin{bmatrix} \cos\alpha & -\sin\alpha & 0 \\ \sin\alpha & \cos\alpha & 0 \\ 0 & 0 & 1 \end{bmatrix}$$

are in fact orthogonal since their product is:

$$\begin{bmatrix} \cos^2\alpha + \sin^2\alpha, & -\cos\alpha \sin\alpha + \cos\alpha \sin\alpha, & 0 \\ -\cos\alpha \sin\alpha + \cos\alpha \sin\alpha, & \cos^2\alpha + \sin^2\alpha, & 0 \\ 0 & 0 & 1 \end{bmatrix} = \begin{bmatrix} 1 & 0 & 0 \\ 0 & 1 & 0 \\ 0 & 0 & 1 \end{bmatrix} = \mathrm{I}$$

Returning now to the question of relating $\boldsymbol{p}'$ and $\boldsymbol{E}'$, we have:

$$\left.\begin{aligned} p'_x &= a_x \cos\alpha(\cos\alpha E'_x - \sin\alpha E'_y) + a_y \sin\alpha(\sin\alpha E'_x + \cos\alpha E'_y) \\ p'_y &= -a_x \sin\alpha(\cos\alpha E'_x - \sin\alpha E'_y) + a_y \cos\alpha(\sin\alpha E'_x + \cos\alpha E'_y) \\ p'_z &= \qquad 0 \qquad\qquad 0 \qquad + a_z E'_z \end{aligned}\right\}$$

or:

$$\boldsymbol{p}' = \mathrm{A}'\boldsymbol{E}'$$

A′ being the matrix:

$$\begin{bmatrix} a_x \cos^2\alpha + a_y \sin^2\alpha & (a_y - a_x) \sin\alpha \cos\alpha & 0 \\ (a_y - a_x) \sin\alpha \cos\alpha & a_x \sin^2\alpha + a_y \cos^2\alpha & 0 \\ 0 & 0 & 1 \end{bmatrix}$$

So we see that p'_x is no longer a simple scalar times E'_x; it also involves E'_y. Thus the relationship between $\boldsymbol{E}$ and $\boldsymbol{p}$, if the body is not isotropic, is *in general* a matrix relationship. The simpler relationships:

$$p_x = a_x E_x, \qquad p_y = a_y E_y, \qquad p_z = a_z E_z$$

hold only in *one special coordinate system* (which in all but the simplest cases cannot be determined a priori). In fact the finding of this coordinate system is one of the central and certainly the most important question associated with any linear relationship of the form $\boldsymbol{y} = \mathrm{A}\boldsymbol{x}$ when $\boldsymbol{y}, \boldsymbol{x}$ are vectors. Unless we know, or fortuitously choose, this coordinate system, the most general linear relationship between physical vectors, such as $\boldsymbol{p}$ and $\boldsymbol{E}$ in the example just given, is provided by the matrix product.

Transformation of coordinates

In many physical problems it is necessary, or highly desirable, to change coordinate systems. If, for example, the scattering of light by a particle is being considered, the plane containing the particle and the direction of the incident and scattered radiation has special significance and evidently certain symmetries exist with respect to this plane; but light being a transverse vibration there is also a special significance to the plane perpendicular to the direction of incidence and the plane perpendicular to the direction of the scattered light. In terms of the incident radiation it is natural to describe the light by say electric field components E_x, E_y in arbitrary x, y directions in the plane transverse to the incident light; but when we come to the actual scattering process it is necessary to consider the components perpendicular to (E_r) and parallel with (E_l) the scattering plane. If the original x-axis made an angle α with the scattering plane, this implies an elementary rotation transformation:

$$\begin{bmatrix} E_r \\ E_l \end{bmatrix} = \begin{bmatrix} \cos\alpha & -\sin\alpha \\ \sin\alpha & \cos\alpha \end{bmatrix} \begin{bmatrix} E_x \\ E_y \end{bmatrix}$$

Following scattering, we obtain scattered light which is related to E_r and E_l by:

$$\begin{bmatrix} E'_r \\ E'_l \end{bmatrix} = \begin{bmatrix} s_{11} & s_{12} \\ s_{21} & s_{22} \end{bmatrix} \begin{bmatrix} E_r \\ E_l \end{bmatrix}$$

Finally it is generally necessary to transform back to ξ, η coordinates in the plane transverse to the scattered direction. If we are dealing with three-dimensional scattering the ξ, η directions will usually differ from the x, y directions — this will be the case, for example, if we envisage the particle to be at the center of a unit sphere and define (x, y) and (ξ, η) to coincide with the meridional (i.e. north-south) direction and azimuthal (east-west) directions respectively at those points on the unit sphere where the incident light enters and the scattered light emerges. The final form of the relation between scattered and incident electric field amplitudes is therefore:

$$\begin{bmatrix} E_\xi \\ E_\eta \end{bmatrix} = \begin{bmatrix} \cos\beta & -\sin\beta \\ \sin\beta & \cos\beta \end{bmatrix} \begin{bmatrix} s_{11} & s_{12} \\ s_{21} & s_{22} \end{bmatrix} \begin{bmatrix} \cos\alpha & -\sin\alpha \\ \sin\alpha & \cos\alpha \end{bmatrix} \begin{bmatrix} E_x \\ E_y \end{bmatrix}$$

and it is here physically necessary to define the matrix-vector product as it was defined in Chapter 3, and as we have seen, the matrix-matrix product definition follows logically from the definition of the matrix-vector product.

In this section we have illustrated the physical and geometric reasonableness of the definitions given in the algebraic discussion. The reader by now may be convinced that there are common-sense grounds for the multiplication rules of matrix algebra. He probably also realizes that processes such as rotation of orthogonal axes, for example, by their very nature cannot destroy any information: a point is located equally well in any rectangular coordinate system. Hence the reverse process must be as stable as the forward process and the calculation of the inverse matrix, if needed, should not be especially troublesome, since it is equivalent simply to a rotation through a negative angle. But all matrices are not rotation matrices, as we shall soon see.

4.2. NORMS, LENGTH AND DISTANCE

The concept of length comes naturally in geometry and it is natural to carry it over into algebra in the case of vectors of arbitrary dimension. The length of the vector with components x, y, z is $\sqrt{x^2 + y^2 + z^2}$. (The length of a vector with spherical coordinates r, θ, ϕ, however, is *not* $\sqrt{r^2 + \theta^2 + \phi^2}$.) If we are dealing with algebraic vectors it is natural to think of the elements as being x-, y-, z-components of a geometric vector, and it is only a small step beyond that to a generalization in which the length of a vector $\boldsymbol{v}$ with n elements is $\sqrt{v_1^2 + v_2^2 + v_3^2 + \ldots + v_n^2}$. Recalling the definition of the matrix product and treating the vector $\boldsymbol{v}$ as a $N \times 1$ matrix and its transpose $\boldsymbol{v}^*$ as a $1 \times N$ matrix, we see that the length can also be considered as follows:

$$(\text{length of } \boldsymbol{v})^2 = \boldsymbol{v}^* \boldsymbol{v}$$

scalar or 1×1 matrix $\qquad (1 \times N)(N \times 1)$

This kind of product of course is the "dot product", "scalar product" or "inner product" $\boldsymbol{v} \cdot \boldsymbol{v}$ of geometric vector analysis.

A unit vector (i.e. a vector of unit length) can obviously be obtained from any non-trivial vector merely by dividing by its length. This process is called normalization and when the unitary nature of a unit vector is to be emphasized we will use a cap symbol above the vector symbol, so if some $\boldsymbol{v}$ has $\boldsymbol{v}^* \boldsymbol{v} = |\boldsymbol{v}|^2$, the vector represented by $\hat{\boldsymbol{v}}$ will be $\boldsymbol{v}/|\boldsymbol{v}|$, a unit vector. A set of unit vectors which are mutually orthogonal are known as an "orthonormal" set.

If now we form the dot product of two vectors $\boldsymbol{v}$ and $\boldsymbol{u}$, which can be written $\boldsymbol{v}^* \boldsymbol{u}$ or $\boldsymbol{u}^* \boldsymbol{v}$, we find a scalar which in general is not equal in magnitude to the product of the lengths of the individual vectors. For elementary vectors of dimension 2, this is readily proved:

$$|\boldsymbol{v}| = \text{length of } \boldsymbol{v} = \sqrt{v_1^2 + v_2^2}$$

$$|\boldsymbol{u}| = \text{length of } \boldsymbol{u} = \sqrt{u_1^2 + u_2^2}$$

Scalar product $\boldsymbol{v}^* \boldsymbol{u} = v_1 u_1 + v_2 u_2$.

To clear the square roots, we compare the squares of the lengths, i.e.:

$$(v_1^2 + v_2^2)(u_1^2 + u_2^2) \qquad \text{with } (v_1 u_1 + v_2 u_2)^2$$

or

$$v_1^2 u_1^2 + v_2^2 u_2^2 + v_1^2 u_2^2 + v_2^2 u_1^2 \qquad \text{with } v_1^2 u_1^2 + v_2^2 u_2^2 + 2 v_1 v_2 u_1 u_2$$

or

$$(v_1 u_2)^2 + (v_2 u_1)^2 \qquad \text{with } 2(v_1 u_2)(v_2 u_1)$$

or

$$a^2 + b^2 \qquad \text{with } 2ab$$

but $(a^2 + b^2 - 2ab) = (a - b)^2$ is never negative, so the left side is never less than the right side, so:

$$|v|^2 |u|^2 \geqslant (v^* u)^2 = (v \cdot u)^2$$

or

$$|v||u| \geqslant v^* u \qquad [4.1a]$$

This result can be derived in a similar way for vectors of arbitrary dimension. It is a result of especial importance in applied mathematics and is known as Schwarz's inequality. The result can be written in terms of summation:

$$\sum_i x_i^2 \sum_i y_i^2 \geqslant (\sum_i x_i y_i)^2 \qquad [4.1b]$$

It may be noted that the generalized length of a vector and the expression $\int_a^b f^2(x)\, dx$ both in a sense measure the magnitude of the vector or function. There is a very close parallel between functions and vectors and the parallelism is accentuated in numerical analysis where integrals are almost always replaced by summation for their numerical evaluation. This process, known as "numerical quadrature" replaces the function $f(x)$ by a vector containing the values $f_1, f_2, ..., f_N$ of $f(x)$ at $x = x_1, x_2, ..., x_N$ and replaces the integration $\int_a^b f(x)\, dx$ by a sum of the form $w \, \Sigma_i f_i$ or $\Sigma_i w_i f_i$. The essential equivalence of functions and vectors leads to the use of terms such as length for $\sqrt{\int_a^b f^2(x)\, dx}$, dot product or scalar product for $\int_a^b f(x)\, g(x)\, dx$, parallel for functions which differ only by a scalar factor, and so on. Just as three-dimensional space contains all three-dimensional geometric vectors, N-dimensional *vector space* contains all N-dimensional vectors and a *function space* may contain all functions of a prescribed character (e.g. continuous and differentiable in (a, b), of bounded variation in (a, b), or whatever).

The measures of magnitude $|...|$ and $|...|^2$ whether applied to vectors or functions are used so often that names are desirable for them. In what follows we will use "length" for $|...|$ and "square norm" for $|...|^2$; Courant and Hilbert in their text define the latter as the norm, but there are in the literature other "norms" and for clarity we will emphasize which norm is intended here by speaking of the square norm. For the sake of clarity we show in Table 4.1 explicit expressions for the various quantities and properties which arise as a result of this geometric picture of functions and vectors.

4.3. ORTHOGONALITY

The concept of orthogonality is of fundamental importance in many fields, just as orthogonal axes are of fundamental importance in geometry. Some arbitrary pair of vectors $\boldsymbol{f}$ and $\boldsymbol{g}$ will in general be neither orthogonal ($\boldsymbol{f}^* \boldsymbol{g} = 0$) nor parallel (differing only by a scalar factor — which gives $|\boldsymbol{f}^* \boldsymbol{g}| =$

TABLE 4.1

Notation and mathematical definition of the various quantities and properties of vectors and functions

Quantity or property	Notation	Mathematical definition
For vectors (N-dimensional):		
Inner product Dot product Scalar product	f^*g or $g^*f, f \cdot g, (f, g)$	$\sum_{i=1}^{N} f_i g_i$
Square norm	$\|f\|^2, f^*f, f \cdot f, f^2$	$\sum_{i=1}^{N} f_i^2$
Length	$\|f\|, f$	$\sqrt{\sum_{i=1}^{N} f_i^2}$
Orthogonality	$f \cdot g = 0, f^*g = 0$	$\sum_i f_i g_i = 0$
Normalization	$f^*f = f \cdot f = \|f\|^2 = 1$	$\sum_i f_i^2 = 1$
For functions in the interval (a, b):		
Inner product	$f \cdot g, (f, g)$	$\int_a^b f(x)\, g(x)\, dx$
Square norm	$\|f\|^2$ or $\|f(x)\|^2$	$\int_a^b f^2(x)\, dx$
Length	$\|f\|$ or $\|f(x)\|$	$\sqrt{\int_a^b f^2(x)\, dx}$
Orthogonality	$f \cdot g = (f, g) = 0$	$\int_a^b f(x)\, g(x)\, dx = 0$
Normalization	$f \cdot f = (f, f) = 1$	$\int_a^b f^2(x)\, dx = 1$

$|\boldsymbol{f}||\boldsymbol{g}|$), just as an arbitrary pair of directions will in general subtend an angle θ which is neither zero nor 90°. A vector $\boldsymbol{g}$ derived from $\boldsymbol{f}$ by scalar multiplication alone is in a sense not really a "new" or "different" vector, but merely a restatement of $\boldsymbol{f}$. To be fundamentally different in a vectorial sense

f^*g must not vanish. If g is arbitrary it will in general contain a component proportional to f and another which we can readily show is orthogonal to f. It will be convenient to assume that f is normalized, i.e. that $|f| = 1$ (this can always be brought about by dividing an arbitrary f by its length). Writing:

$$g = \alpha f + \beta u$$

where α, β are to be determined, and u is required to be orthogonal to f and normalized, we can find α and β readily:

$$f^* g = \alpha \qquad \text{(since } f^* u = 0\text{)}$$

$$u^* g = \beta$$

so

$$\beta u = g - \alpha f = g - (f^* g) f$$

There is thus a unique vector u (and it is given by the last relationship), *unless* either of the following circumstances occur:

$$g = (f^* g) f$$

$$\beta = 0$$

Consider first the relationship $g = (f^*g) f$; it defines g as being merely f multiplied by the scalar f^*g, and will hold only when g happens to be in the same direction as f and not independent thereof in a vectorial sense (i.e. if we normalize f and g we get the same unit vector). The other relationship, $\beta = 0$, gives $g = \alpha f$, which again will only happen if g and f happen to be in the same direction and not truly independent. The two relationships $g = (f^*g) f$ and $\beta = 0$ are therefore equivalent.

The Schmidt process

We have just shown that it is possible from two vectors f and g to construct a pair of linear combinations of the two vectors which are mutually orthogonal, provided only that the vectors are not scalar multiples of each other. Note, however, that this process is not unique — we proceeded from f to g in that order, getting from g a vector $g - (f^*g)f$ which was orthogonal to f. We could equally well have started with g and got from f a vector orthogonal to g.

We can continue to apply this procedure to an arbitrary set of vectors, $v_1, v_2, \ldots, v_n$ to produce a set of orthogonal and normalized (orthonormal) vectors $\hat{u}_1, \hat{u}_2, \hat{u}_3, \ldots$. If none of the vectors $v_1, v_2, \ldots, v_n$ can be written as a linear combination of the others, we will be able to produce as many orthogonal vectors $\hat{u}_1, \hat{u}_2, \ldots,$ as there are v's. This process is as follows:

(1) take v_1, calculate $|v_1|$ and normalize to obtain $\hat{u}_1 = v_1/|v_1|$,

(2) take $\boldsymbol{v}_2$, calculate $\boldsymbol{v}_2 - (\hat{\boldsymbol{u}}_1 \cdot \boldsymbol{v}_2)\,\hat{\boldsymbol{u}}_1$, which is orthogonal to $\hat{\boldsymbol{u}}_1$, as can be seen by forming the scalar product with $\hat{\boldsymbol{u}}_1$,
(3) calling this $\boldsymbol{u}_2$, calculate $|\boldsymbol{u}_2|$ and form the normalized vector $\hat{\boldsymbol{u}}_2 = \boldsymbol{u}_2/|\boldsymbol{u}_2|$,
(4) take $\boldsymbol{v}_3$ and calculate $\boldsymbol{v}_3 - (\hat{\boldsymbol{u}}_1 \cdot \boldsymbol{v}_3)\,\hat{\boldsymbol{u}}_1 - (\hat{\boldsymbol{u}}_2 \cdot \boldsymbol{v}_3)\,\hat{\boldsymbol{u}}_2$, which is orthogonal to both $\boldsymbol{u}_1$ and $\boldsymbol{u}_2$,
(5) calling this $\boldsymbol{u}_3$, normalize to obtain $\hat{\boldsymbol{u}}_3$,
(6) continue in like manner. If any vector $\boldsymbol{v}_m$ is a linear combination of its predecessors it will give a zero vector for $\boldsymbol{u}_m = \boldsymbol{v}_m - (\hat{\boldsymbol{u}}_1 \cdot \boldsymbol{v}_m)\,\hat{\boldsymbol{u}}_1 - (\hat{\boldsymbol{u}}_2 \cdot \boldsymbol{v}_m)\,\hat{\boldsymbol{u}}_2 \ldots - (\hat{\boldsymbol{u}}_{m-1} \cdot \boldsymbol{v}_m)\,\hat{\boldsymbol{u}}_{m-1}$ and can be deleted.

This process is known as the Schmidt process and has considerable theoretical and practical significance in many fields. Some applications of the Schmidt process will be encountered later and hopefully the merits and unique advantages of orthogonal representation in vector- and function-space will also become increasingly obvious as we continually apply them.

4.4. GEOMETRICAL VIEW OF MATRIX OPERATIONS

A matrix operates (premultiplies) a vector, giving another vector. If the operation does not change the length of the vector, the matrix is a special kind of matrix, often known as a unitary matrix — the rotation matrices encountered earlier were of this character. Generally the length of the new vector will depend on the direction, as well as the length, of the original vector, i.e. $|\mathbf{A}\boldsymbol{v}|/|\boldsymbol{v}|$ will depend upon $\boldsymbol{v}$ in the general case; if $|\mathbf{A}\boldsymbol{v}|/|\boldsymbol{v}|$ is constant $\mathbf{A}$ can be factored into a scalar and a unitary matrix $\mathbf{A}'$ for which $|\mathbf{A}'\boldsymbol{v}| = |\boldsymbol{v}|$. The identity matrix $\mathbf{I}$ is of course a unitary matrix.

Another possibility is that $\mathbf{A}\boldsymbol{v}$ is not $\boldsymbol{v}$ or a scalar multiple thereof, but that $\mathbf{A}^*\mathbf{A}\boldsymbol{v}$ is. If that holds for arbitrary $\boldsymbol{v}$, $\mathbf{A}^*\mathbf{A}$ must be a diagonal matrix. The relationship:

$$\mathbf{A}^*\mathbf{A} = \text{(diagonal matrix)}$$

defines the *orthogonality* of $\mathbf{A}$. For $\mathbf{A}$ to be orthogonal it is evidently sufficient that the dot products of the columns of A, taken in pairs, vanish (except of course when a column is dotted with itself). If $\mathbf{A}^*\mathbf{A}$ happens to be the special diagonal matrix $\mathbf{I}$, $\mathbf{A}$ is said to be normalized as well as orthogonal, or simply "orthonormal".

Orthogonal matrices can be very simple in form. Consider the matrix:

$$\mathbf{R} = \begin{bmatrix} 1 & 0 & 0 & 0 & 0 \\ 0 & a & 0 & c & 0 \\ 0 & 0 & 1 & 0 & 0 \\ 0 & b & 0 & d & 0 \\ 0 & 0 & 0 & 0 & 1 \end{bmatrix}$$

Inspection shows that the first, third and fifth columns are obviously orthogonal with respect to one another, and also with respect to the second and fourth column. **R** will therefore be orthogonal if the second and fourth columns are mutually orthogonal, i.e. if:

$$ac + bd = 0$$

If we agree to normalize these columns we have also:

$$a^2 + b^2 = 1$$

$$c^2 + d^2 = 1$$

an obvious choice which satisfies the latter two identities is $a = \cos\alpha$, $b = \sin\alpha$; $d = \cos\beta$, $c = \sin\beta$; the orthogonality condition then becomes:

$$\cos\alpha \sin\beta + \sin\alpha \cos\beta = 0$$

or:

$$\sin(\alpha + \beta) = 0,$$

which is satisfied by $\beta = -\alpha$. So the matrix:

$$\mathbf{R}_{mn} = \begin{bmatrix} 1 & & & & & & & & & & & & \\ & 1 & & & & & & & & & & & \\ & & 1 & & & & & & & & & & \\ & & & \cos\alpha & & & & & & -\sin\alpha & & & \\ & & & & 1 & & & & & \cdot & & & \\ & & & & & 1 & & & & \cdot & & & \\ & & & & & & 1 & & & 0 & & & \\ & & & & & & & 1 & & \cdot & & & \\ & & & & & & & & 1 & & & & \\ & & & \sin\alpha & & & & & & \cos\alpha & & & \\ & & & & & & & & & & \cdot & & \\ & & & & & & & & & & & \cdot & \\ & & & & & & & & & & & & 1 \end{bmatrix} = \|r_{ij}\|$$

$r_{ii} = 1$ $(i \neq m, \neq n)$; $r_{ij} = 0$ $(i \neq j;\ i,j \neq m;\ i,j \neq n)$; $r_{mm} = r_{nn} = \cos\alpha$; $r_{mn} = -r_{nm} = \sin\alpha$.

(off-diagonal elements not written being zero) is orthogonal and normal, or "orthonormal". Note that $\boldsymbol{v}$ and the product $\mathbf{R}\boldsymbol{v}$ have equal lengths in that case since:

$$(\text{length of } v)^2 = v^* v$$

$$(\text{length of } \mathbf{R}v)^2 = (\mathbf{R}v)^* (\mathbf{R}v)$$

$$= v^* \mathbf{R}^* \mathbf{R} v$$

$$= v^* \mathbf{I} v \qquad = v^* v$$

So that:

$|\mathbf{R}v| = |v|$ *when* **R** *is orthogonal and normal*

Thus $\mathbf{R}v$ represents a "rotation" of the vector v in which its length is preserved. The phrase "unitary transformation" may also be encountered. The product **RA** is not generally symmetric when **A** is symmetric, but the product $\mathbf{RAR}^*$ does preserve symmetry. This kind of product has a special importance, as will be seen directly.

Suppose now **R** is of the form shown above, with zeros off the diagonal and unity along the diagonal except in the mth row and nth column where $\cos\alpha$ appears in the diagonal (m, m) and (n, n) positions, $\sin\alpha$ in the (m, n) position and $-\sin\alpha$ in the (n, m) position. Suppose also that **A** is a symmetric matrix, so that $a_{mn} = a_{nm}$. Premultiplication of **A** by **R** amounts to multiplying the mth row of **A** by $\cos\alpha$, multiplying the nth row by $-\sin\alpha$, adding the results to obtain a new mth row, while a new nth row is formed by adding $\sin\alpha$ times the original mth row and $\cos\alpha$ times the original nth row. The remaining rows and columns are left unchanged, so the effect of operations involving **R** can be described in terms of 2×2 matrices. $\mathbf{RAR}^*$ is obtained by row operations on $\mathbf{AR}^*$ (which is obtained from **A** by column operations):

$$\begin{matrix} \mathbf{R} & \mathbf{A} & \mathbf{R}^* \end{matrix}$$

$$= \begin{bmatrix} \cos\alpha & -\sin\alpha \\ \sin\alpha & \cos\alpha \end{bmatrix} \begin{bmatrix} a_{mm} & a_{mn} \\ a_{mn} & a_{nn} \end{bmatrix} \begin{bmatrix} \cos\alpha & \sin\alpha \\ -\sin\alpha & \cos\alpha \end{bmatrix}$$

$$= \begin{bmatrix} \cos\alpha & -\sin\alpha \\ \sin\alpha & \cos\alpha \end{bmatrix} \begin{bmatrix} a_{mm}\cos\alpha - a_{mn}\sin\alpha & a_{mm}\sin\alpha + a_{mn}\cos\alpha \\ a_{mn}\cos\alpha - a_{nn}\sin\alpha & a_{mn}\sin\alpha + a_{nn}\cos\alpha \end{bmatrix}$$

So its elements are

$(1, 1)$: $a_{mm}\cos^2\alpha - a_{mn}\cos\alpha\sin\alpha - a_{mn}\cos\alpha\sin\alpha + a_{nn}\sin^2\alpha$

$(1, 2)$: $a_{mm}\sin\alpha\cos\alpha + a_{mn}\cos^2\alpha - a_{mn}\sin^2\alpha - a_{nn}\cos\alpha\sin\alpha$

$(2, 2)$: $a_{mm}\sin^2\alpha + a_{mn}\cos\alpha\sin\alpha + a_{mn}\cos\alpha\sin\alpha + a_{nn}\cos^2\alpha$

$(2, 1)$: $a_{mm}\cos\alpha\sin\alpha - a_{mn}\sin^2\alpha + a_{mn}\cos^2\alpha - a_{nn}\cos\alpha\sin\alpha$

The equal off-diagonal elements in the (1, 2) and (2, 1) positions will be zero if:

$$a_{mm} \cos \alpha \sin \alpha + a_{mn}(\cos^2\alpha - \sin^2\alpha) - a_{nn} \cos \alpha \sin \alpha = 0$$

or

$$(a_{nn} - a_{mm}) \sin 2\alpha = 2a_{mn} \cos 2\alpha$$

or

$$\tan 2\alpha = 2a_{mn}/(a_{nn} - a_{mm}) \qquad [4.2]$$

Any selected off-diagonal element can in this way be made zero in $\mathbf{RAR}^*$ by this choice for α. Since we have made use of $a_{mn} = a_{nm}$, it is applicable only when $\mathbf{A}$ is symmetric.

By a repetition of this procedure (fairly easily coded for a computer) it is evident that we can eventually reduce all off-diagonal elements towards zero. This is equivalent to finding a sequence of $\mathbf{R}$'s, $\mathbf{R}_1, \mathbf{R}_2, \ldots, \mathbf{R}_N$, such that:

$$\mathbf{R}_N\mathbf{R}_{N-1} \ldots \mathbf{R}_2\mathbf{R}_1\mathbf{A}\mathbf{R}_1^*\mathbf{R}_2^* \ldots \mathbf{R}_N^* = \mathbf{D}$$

where $\mathbf{D}$ is diagonal or almost diagonal. Note that the number N of iterations required to reduce all off-diagonal elements below some specified level cannot be predicted a priori, since the element which is made zero at some stage does not necessarily remain so through later iterations. However, the method is a practical and popular method; originally due to Jacobi, it was rediscovered by von Neumann in the early days of modern electronic computers when practical methods for so-called "diagonalization" were being sought.

The product of orthogonal matrices $\mathbf{R}_N\mathbf{R}_{N-1} \ldots, \mathbf{R}_1$ is itself an orthogonal matrix, for we have:

$$(\mathbf{R}_N\mathbf{R}_{N-1} \ldots \mathbf{R}_1)^*(\mathbf{R}_N\mathbf{R}_{N-1} \ldots \mathbf{R}_1) = \mathbf{R}_1^*\mathbf{R}_2^* \ldots \mathbf{R}_{N-1}^*\mathbf{R}_N^*\mathbf{R}_N\mathbf{R}_{N-1} \ldots \mathbf{R}_2\mathbf{R}_1$$

The innermost product pair gives $\mathbf{I}$, as do succeeding innermost pairs, so that the above is itself equivalent to $\mathbf{I}$, and $(\mathbf{R}_N\mathbf{R}_{N-1} \ldots, \mathbf{R}_1)$ is an orthogonal matrix. If we denote this simply by $\mathbf{R}$, we have:

$$\mathbf{RAR}^* = \mathbf{D} \qquad [4.3a]$$

Multiplying by $\mathbf{R}^*$ and noting that $\mathbf{R}^*\mathbf{R} = \mathbf{I}$, it follows that:

$$\mathbf{AR}^* = \mathbf{R}^*\mathbf{D}$$

and

$$\mathbf{A} = \mathbf{R}^*\mathbf{DR} \qquad [4.3b]$$

Thus $\mathbf{A}$ has been "factored" into the product of an orthogonal matrix $\mathbf{R}^*$, a diagonal matrix $\mathbf{D}$ and the "paired" orthogonal matrix $\mathbf{R}$. It will be seen in the next section that this solves the problem of obtaining vector(s) such that $\mathbf{A}\boldsymbol{x} = \lambda\boldsymbol{x}$.

4.5. EIGENVALUES AND EIGENVECTORS

The product $\mathbf{A}\boldsymbol{x}$ is a vector $\boldsymbol{y}$ which in general has no simple relationship to the vector $\boldsymbol{x}$. When $\mathbf{A}$ is the identity matrix or a multiple thereof, formation of $\boldsymbol{y}$ requires only the multiplication of $\boldsymbol{x}$ (element by element) by a single number. For example:

$$\begin{bmatrix} 2 & 0 & 0 \\ 0 & 2 & 0 \\ 0 & 0 & 2 \end{bmatrix} \begin{bmatrix} x_1 \\ x_2 \\ x_3 \end{bmatrix} = \begin{bmatrix} 2x_1 \\ 2x_2 \\ 2x_3 \end{bmatrix}$$

It is natural to enquire whether for arbitrary $\mathbf{A}$ there is a choice or choices of $\boldsymbol{x}$ which make $\mathbf{A}\boldsymbol{x}$ a simple scalar multiple of $\boldsymbol{x}$, so that:

$$\mathbf{A}\boldsymbol{x} = \lambda\boldsymbol{x} \qquad [4.4]$$

Clearly $\mathbf{A}$ must be square (equal number of rows and colums) for this to be possible. The following numerical exercise shows that there can be such a vector (it does not of course prove that there must be). The exercise, though simple, is hardly possible without the aid of some sort of computer; it consists of repeated premultiplication of an arbitrary vector $\boldsymbol{x}$ by $\mathbf{A}$, i.e. there are calculated:

$$\boldsymbol{x}_1 = \mathbf{A}\boldsymbol{x}, \qquad \boldsymbol{x}_2 = \mathbf{A}\boldsymbol{x}_1, \qquad \ldots\ \boldsymbol{x}_n = \mathbf{A}\boldsymbol{x}_{n-1}$$

and so on. The result of course is equivalent to calculating $\mathbf{A}^n\boldsymbol{x}$ but repeated matrix-vector multiplications are very much faster than repeated matrix multiplications. For $\mathbf{A}$ and $\boldsymbol{x}$ the following values were used:

$$\mathbf{A} = \begin{bmatrix} 1.0 & 0.8 & 0.6 & 0.4 \\ 0.8 & 2.0 & 1.0 & 0.6 \\ 0.6 & 1.0 & 2.0 & 0.5 \\ 0.4 & 0.6 & 0.5 & 1.0 \end{bmatrix} \qquad \boldsymbol{x} = \begin{bmatrix} 1 \\ 1 \\ 1 \\ 1 \end{bmatrix}$$

The results are laid out from left to right below:

$$\begin{bmatrix} 1 \\ 1 \\ 1 \\ 1 \end{bmatrix} \begin{bmatrix} 2.8 \\ 4.4 \\ 4.1 \\ 2.5 \end{bmatrix} \begin{bmatrix} 9.78 \\ 16.64 \\ 15.53 \\ 8.31 \end{bmatrix} \begin{bmatrix} 35.734 \\ 61.620 \\ 57.723 \\ 29.971 \end{bmatrix} \begin{bmatrix} 131.652 \\ 227.533 \\ 213.492 \\ 110.098 \end{bmatrix} \begin{bmatrix} 485.813 \\ 839.938 \\ 788.557 \\ 406.025 \end{bmatrix} \begin{bmatrix} 1793.307 \\ 3100.698 \\ 2911.552 \\ 1498.59 \end{bmatrix} \begin{bmatrix} 6620.23 \\ 11446.75 \\ 10749.08 \\ 5532.11 \end{bmatrix}$$

Inspection of the above list of vectors shows that an almost constant ratio, element by element, is soon brought about between $\boldsymbol{x}_n$ and $\boldsymbol{x}_{n-1}$. For $\boldsymbol{x}_8$ and $\boldsymbol{x}_7$ above the ratios are:

3.69163, 3.69167, 3.69187, 3.69154

Thus already a vector has been obtained which to high accuracy exhibits the behavior $\mathbf{A}\boldsymbol{x} = \lambda \boldsymbol{x}$. At this stage it should be pointed out that the equation $\mathbf{A}\boldsymbol{x} = \lambda\boldsymbol{x}$ is indeterminate to the extent of a multiplicative factor; if $\boldsymbol{x}$ satisfies the equation so also does $2\boldsymbol{x}$, $-\frac{1}{3}\boldsymbol{x}$, in fact any multiple of $\boldsymbol{x}$. To make $\boldsymbol{x}$ unique it must be normalized in some way; a convenient and meaningful way to do so is to stipulate that $\boldsymbol{x}$ is a unit vector, in the sense that $|\boldsymbol{x}|^2 = \Sigma_k x_k^2 = 1$. If we normalize $\boldsymbol{x}_8$ and $\boldsymbol{x}_9$ above we obtain:

$$\hat{\boldsymbol{x}}_8 = \begin{bmatrix} 0.369502 \\ 0.638889 \\ 0.599960 \\ 0.308767 \end{bmatrix}; \qquad \hat{\boldsymbol{x}}_9 = \begin{bmatrix} 0.369501 \\ 0.638888 \\ 0.599962 \\ 0.308767 \end{bmatrix}$$

These are therefore close approximations to a vector $\boldsymbol{u}$ which has the property:

$$\mathrm{A}\boldsymbol{u} = \lambda\boldsymbol{u}$$

and λ is approximated closely by the ratios listed above (i.e. λ is 3.691...).

When it satisfies a relationship of this sort a vector is designated an *eigenvector* of the matrix $\mathbf{A}$ and the scalar multiplier λ is known as the *eigenvalue*. We have by the above numerical procedure obtained a single eigenvector and eigenvalue, but in general an $N \times N$ matrix will possess N eigenvectors, each of which is paired with an eigenvalue. (The expressions "characteristic vectors" and "characteristic values" are also encountered.) That there will in general be N eigenvectors and eigenvalues follows from the fact that the matrix-vector equation is a system of N scalar equations in the N unknown components of $\boldsymbol{u}$ and the single additional unknown λ; the N components of $\boldsymbol{u}$ are connected by the normalization condition $\Sigma_i u_k^2 = 1$.

Determination of λ, $\boldsymbol{u}$ such that $A\boldsymbol{u} = \lambda\boldsymbol{u}$, A being real and symmetric

Earlier it was found that there is a matrix $\mathbf{R}^*$ such that (equation [4.3]):

$$\mathbf{AR}^* = \mathbf{R}^*\mathbf{D}$$

with $\mathbf{D}$ a diagonal matrix. Now postmultiplication of a matrix $\mathbf{R}^*$ by a diagonal $\mathbf{D}$, gives a matrix with (i, j) element $r_{ij}^* d_{jj}$, since every other d_{kj} in the sum $\Sigma_k\ r_{ik}^* d_{kj}$ vanishes. Thus the product $\mathbf{R}^*\mathbf{D}$ is simply $\mathbf{R}^*$ modified by multiplying its first column by d_{11}, its second column by d_{22}, and so on. But the first column in $\mathbf{AR}^*$ is the matrix-vector product of $\mathbf{A}$ with the first column of $\mathbf{R}^*$. Thus if $\boldsymbol{u}_1$ is the first column of $\mathbf{R}^*$, $\boldsymbol{u}_2$ the second column, and so on, then:

$$\mathbf{A}u_1 = d_{11}u_1$$

$$\mathbf{A}u_2 = d_{22}u_2$$

.

.

$$\mathbf{A}u_n = d_{nn}u_n$$

So when the Jacobi diagonalization has been completed, up to N vectors and scalar multipliers are at hand each of which satisfy the eigenvalue relationship:

$$\mathbf{A}u = \lambda u$$

Note that this diagonalization procedure depends on the symmetry of **A**; its operations are all real, so that the eigenvalues and eigenvectors of a real symmetric matrix are real.

The scalars d_{11}, d_{22}, . . . etc., are the eigenvalues and the $\boldsymbol{u}$'s are the eigenvectors of the matrix **A**. Their order is immaterial but it will be convenient to take them in order of the magnitude of the eigenvalues. The matrix **D** will from now on be called Λ since its elements are the eigenvalues λ_1, λ_2, etc.

We will not discuss here in detail the special situations where an $(N \times N)$ symmetric matrix does not give N distinct eigenvalues and eigenvectors. The interested reader will find discussions in texts on matrix algebra and the algebraic eigenvalue problem. Obviously the eigenvector can be any vector whatever when an eigenvalue is zero; if two eigenvalues are equal, so that:

$$\mathbf{A}u_{\mathrm{a}} = \lambda u_{\mathrm{a}}$$

$$\mathbf{A}u_{\mathrm{b}} = \lambda u_{\mathrm{b}}$$

then the eigenvectors are degenerate in that any linear combination $\xi u_{\mathrm{a}} + \eta u_{\mathrm{b}}$ also satisfy the identity:

$$\mathbf{A}(\xi u_{\mathrm{a}} + \eta u_{\mathrm{b}}) = \lambda(\xi u_{\mathrm{a}} + \eta u_{\mathrm{b}})$$

Orthogonality of eigenvectors

The procedure described above automatically ensures that the eigenvectors (the columns of $\mathbf{R}^*$) are orthogonal. We have, however, assumed symmetry in **A**. A more usual procedure of handling eigenvalue/eigenvector analysis is to look at the basic identity:

$$\mathbf{A}u = \lambda u$$

or its component form:

$$\left.\begin{aligned} &(a_{11}-\lambda)\,u_1 + a_{12}u_2 + \ldots + a_{1N}u_N = 0 \\ &a_{21}u_1 + (a_{22}-\lambda)\,u_2 + \ldots + a_{2N}u_N = 0 \\ &\cdot \\ &\cdot \\ &a_{N1}u_1 + \ldots + (a_{NN}-\lambda)\,u_N = 0 \end{aligned}\right\}$$

This last system of equations is homogenous with respect to the variables $u_1, u_2, \ldots, u_N$ and according to the theory of equations can have a non-trivial solution only when the associated determinant vanishes. The coefficients on the left side are elements of the matrix $(\mathbf{A} - \lambda\mathbf{I})$ and the determinant is $\det(\mathbf{A} - \lambda\mathbf{I})$. For a 2×2 matrix $\mathbf{A}$ the determinant is:

$$\begin{aligned} \det(\mathbf{A}-\lambda\mathbf{I}) &= (a_{11}-\lambda)(a_{22}-\lambda) - a_{12}a_{21} \\ &= \lambda^2 - (a_{11}+a_{22})\,\lambda + a_{11}a_{22} - a_{12}a_{21} \end{aligned}$$

— a quadratic in λ, which has two roots λ_1 and λ_2 (which can be complex). In general the determinant $\det(\mathbf{A} - \lambda\mathbf{I})$ is an N-order polynomial in λ, possessing N roots $\lambda_1, \lambda_2, \ldots, \lambda_N$, some of which can be complex numbers. Once these roots — which are the eigenvalues of the matrix $\mathbf{A}$ — are found, the eigenvectors $\boldsymbol{u}_1, \boldsymbol{u}_2, \ldots, \boldsymbol{u}_N$ follow immediately, these being the solution sets for the unknown array $(u_1, u_2, \ldots, u_N)$; one such solution exists for each eigenvalue, but two eigenvalues may happen to give identical eigenvectors. The equation:

$$\det(\mathbf{A} - \lambda\mathbf{I}) = 0$$

is an N-order polynomial in λ. It is useful to note that the term independent of λ in that polynomial is necessarily equal to $\det(\mathbf{A})$. If $\det(\mathbf{A})$ vanishes, that implies that $\lambda = 0$ is a root of the equation for the eigenvalues, so that $\mathbf{A}$ has at least one eigenvalue which vanishes; such a condition is known as *singularity* and $\mathbf{A}$ in that event is a *singular matrix*. A singular matrix cannot be inverted. If its eigenvalues are all positive $\mathbf{A}$ is said to be *positive definite*.

If $\mathbf{A}$ is symmetric and non-singular the eigenvectors $\boldsymbol{u}_1, \boldsymbol{u}_2, \ldots, \boldsymbol{u}_N$ and the eigenvalue $\lambda_1, \lambda_2, \ldots, \lambda_N$ categorize $\mathbf{A}$ completely. If $\mathbf{A}$ is not symmetric there is a second eigen-equation, that for the transpose $\mathbf{A}^*$:

$$\mathbf{A}^* \boldsymbol{v} = \lambda \boldsymbol{v}$$

Since a determinant is unaffected by transposition, $\mathbf{A}$ and $\mathbf{A}^*$ have the same eigenvalues but the homogeneous system of equations for the eigenvectors is different unless $\mathbf{A}$ is symmetric, so to categorize fully an arbitrary matrix the eigenvalues $\lambda_1, \lambda_2, \ldots, \lambda_N$ and *two* sets of vectors, $\boldsymbol{u}_1, \boldsymbol{u}_2, \ldots, \boldsymbol{u}_N$ and $\boldsymbol{v}_1, \boldsymbol{v}_2, \ldots, \boldsymbol{v}_N$ are involved. For a symmetric matrix $\boldsymbol{u}_1$ and $\boldsymbol{v}_1$, $\boldsymbol{u}_2$ and $\boldsymbol{v}_2$, etc., become identical.

The set of vectors $\boldsymbol{u}_1, \boldsymbol{u}_2, ..., \boldsymbol{u}_N$ can be collected as columns of a single matrix **U** and $\boldsymbol{v}_1, \boldsymbol{v}_2, ..., \boldsymbol{v}_N$ can similarly be used to form a matrix **V**. The fundamental equations can be written:

$$\mathbf{AU} = \mathbf{U}\Lambda \qquad [4.5a]$$

if the eigenvalues $\lambda_1, \lambda_2, ..., \lambda_N$ are collected into a diagonal matrix Λ — i.e. Λ is:

$$\begin{bmatrix} \lambda_1 & 0 & 0 & \cdots & 0 \\ 0 & \lambda_2 & 0 & \cdots & 0 \\ 0 & 0 & \lambda_3 & \cdots & 0 \\ . & . & . & \cdots & . \\ . & . & . & \cdots & . \\ 0 & 0 & 0 & \cdots & \lambda_N \end{bmatrix}$$

Λ appears as a *post*multiplier of **U** in the above expression because that leads to the first *column* of **U** being multiplied by λ_1, the second by λ_2, and so on. We have in fact for the (i, j) element of the product $\mathbf{U}\Lambda$ the expression:

$$\sum_j u_{ij}\lambda_j$$

so the first column of $\mathbf{U}\Lambda$ is:

$$\begin{bmatrix} u_{11}\lambda_1 \\ u_{21}\lambda_1 \\ u_{31}\lambda_1 \\ . \\ . \\ u_{N1}\lambda_1 \end{bmatrix}$$

Equation [4.5a] is therefore completely equivalent to the set of equations implied by [4.4]:

$$\left.\begin{aligned} \mathbf{A}\boldsymbol{u}_1 &= \lambda_1\boldsymbol{u}_1 \\ \mathbf{A}\boldsymbol{u}_2 &= \lambda_2\boldsymbol{u}_2 \\ &. \\ &. \\ \mathbf{A}\boldsymbol{u}_N &= \lambda_N\boldsymbol{u}_N \end{aligned}\right\}$$

For the transpose $\mathbf{A}^*$ the equation:

$$\mathbf{A}^*\mathbf{V} = \mathbf{V}\mathbf{\Lambda} \tag{4.5b}$$

evidently holds.

An important orthogonality property can readily be derived from [4.5a] and [4.5b]. The former is:

$$\mathbf{AU} = \mathbf{U\Lambda}$$

Hence:

$$\mathbf{V}^*\mathbf{AU} = \mathbf{V}^*\mathbf{U\Lambda}$$

(premultiplying by $\mathbf{V}^*$). Transposition of [4.5b] followed by postmultiplication by $\mathbf{U}$ gives:

$$\mathbf{V}^*\mathbf{AU} = \mathbf{\Lambda}\,\mathbf{V}^*\mathbf{U}$$

Hence the matrix $\mathbf{V}^*\mathbf{U}$ gives the same result whether *premultiplied or postmultiplied* by the diagonal matrix $\mathbf{\Lambda}$. Let $\mathbf{X} = ||x_{ij}||$ represent the product $\mathbf{V}^*\mathbf{U}$. What has just been shown is that:

$$\mathbf{\Lambda X} = \mathbf{X\Lambda}$$

In component form, the last equation becomes:

$$\begin{bmatrix} \lambda_1 x_{11} & \lambda_2 x_{12} & \cdots \\ \lambda_1 x_{21} & \lambda_2 x_{22} & \cdots \\ \lambda_1 x_{31} & \lambda_2 x_{32} & \cdots \\ & & \\ . & & \end{bmatrix} = \begin{bmatrix} \lambda_1 x_{11} & \lambda_1 x_{12} & \cdots \\ \lambda_2 x_{21} & \lambda_2 x_{22} & \cdots \\ \lambda_3 x_{31} & \lambda_3 x_{32} & \cdots \\ & & \\ . & & \end{bmatrix}$$

or, in element-by-element form:

$$\left.\begin{array}{l} 0 \cdot x_{11} = 0 \\ (\lambda_2 - \lambda_1)\, x_{12} = 0 \\ . \\ . \\ (\lambda_N - \lambda_1)\, x_{1N} = 0 \\ (\lambda_1 - \lambda_2)\, x_{21} = 0 \\ 0 \cdot x_{22} = 0 \\ (\lambda_2 - \lambda_3)\, x_{23} = 0 \\ . \\ . \\ (\lambda_2 - \lambda_N)\, x_{2N} = 0 \\ . \\ \text{etc.} \end{array}\right\}$$

So only the diagonal elements $x_{11}, x_{22}, \ldots, x_{NN}$ can differ from zero. Hence $\mathbf{V}^*\mathbf{U}$ is a diagonal matrix, or, in terms of the eigenvectors:

$$\boldsymbol{v}_m^* \boldsymbol{u}_n = \boldsymbol{v}_m \cdot \boldsymbol{u}_n = \begin{cases} 0 \ (m \neq n) \\ 1 \ (m = n) \end{cases}$$

Summary of orthogonality properties

We may summarize the orthogonality properties in terms of the eigenvectors $\boldsymbol{u}_1, \boldsymbol{u}_2, \ldots, \boldsymbol{u}_N$ of $\mathbf{A}$ and the eigenvectors $\boldsymbol{v}_1, \boldsymbol{v}_2, \ldots, \boldsymbol{v}_N$ of $\mathbf{A}^*$, or the same properties may be given in terms of the matrices $\mathbf{U}$, $\mathbf{V}$ containing the eigenvectors in their columns. The following relationships hold for arbitrary real matrix $\mathbf{A}$:

$$\left.\begin{array}{lll} \text{(a)} & \boldsymbol{v}_m^* \boldsymbol{u}_n = \boldsymbol{u}_n^* \boldsymbol{v}_m = \boldsymbol{u}_m \cdot \boldsymbol{v}_n = 0 \quad (n \neq m) & \\ \text{(b)} & \mathbf{V}^*\mathbf{U} = \mathbf{I}\,; & \mathbf{U}^*\mathbf{V} = \mathbf{I} \\ \text{(c)} & \mathbf{A}\boldsymbol{u}_n = \lambda_n \boldsymbol{u}_n\,; & \mathbf{A}^* \boldsymbol{v}_n = \lambda_n \boldsymbol{v}_n \\ \text{(d)} & \mathbf{A}\mathbf{U} = \mathbf{U}\boldsymbol{\Lambda}\,; & \mathbf{A}^*\mathbf{V} = \mathbf{V}\boldsymbol{\Lambda} \\ \text{(e)} & \mathbf{V}^*\mathbf{A}\mathbf{U} = \boldsymbol{\Lambda}\,; & \mathbf{U}^*\mathbf{A}^*\mathbf{V} = \boldsymbol{\Lambda} \\ \text{(f)} & \mathbf{A} = \mathbf{U}\boldsymbol{\Lambda}\mathbf{V}^*\,; & \mathbf{A}^* = \mathbf{V}\boldsymbol{\Lambda}\mathbf{U}^* \end{array}\right\} \qquad [4.6]$$

If $\mathbf{A}$ is symmetric (and symmetric matrices are very common in inversion problems) then $\mathbf{U}$ and $\mathbf{V}$ are equivalent, and we have:

$$\left.\begin{array}{ll} \text{(a)} & \boldsymbol{u}_m^* \boldsymbol{u}_n = \boldsymbol{u}_m \cdot \boldsymbol{u}_n = 0 \quad (n \neq m) \\ \text{(b)} & \mathbf{U}^*\mathbf{U} = \mathbf{I} \\ \text{(c)} & \mathbf{U}^*\mathbf{A}\mathbf{U} = \boldsymbol{\Lambda} \\ \text{(d)} & \mathbf{A} = \mathbf{U}\boldsymbol{\Lambda}\mathbf{U}^* \end{array}\right\} \quad \textit{(for symmetric } \Lambda \textit{ only)} \qquad [4.7]$$

It should be noted that formula [4.7d] and [4.6f] directly reconstruct $\mathbf{A}$ from the eigenvectors (contained in the columns of $\mathbf{U}$ when $\mathbf{A}$ is symmetric, or in $\mathbf{U}$ and $\mathbf{V}$ if $\mathbf{A}$ is not symmetric) and eigenvalues (contained in the diagonal matrix $\boldsymbol{\Lambda}$). This is a useful property for checking on the correctness of a set of eigenvectors and eigenvalues and also in very many situations has great analytical value, since writing $\mathbf{U}\boldsymbol{\Lambda}\mathbf{U}^*$ or $\mathbf{U}\boldsymbol{\Lambda}\mathbf{V}^*$ in place of $\mathbf{A}$ in a matrix relationship or operation often clarifies the essential nature of the processes involved.

The Cayley-Hamilton equation

The determinant $\Delta(\lambda)$ is evidently a polynomial of order N in λ and can be satisfied by at most N eigenvalues. The explicit equation obtained when

$\Delta(\lambda) = 0$ is expanded into powers of λ is called the characteristic equation associated with the matrix $\mathbf{A}$. It can be written:

$$\lambda^N + \beta_{N-1}\lambda^{N-1} + \beta_{N-2}\lambda^{N-2} + \dots + \beta_1\lambda + \beta_0 = 0$$

or:

$$(\lambda - \lambda_1)(\lambda - \lambda_2) \dots (\lambda - \lambda_N) = 0$$

if the eigenvalues λ_1, λ_2, etc., are all distinct.

If on the left side in the first equation we insert where λ^m appears the matrix $\mathbf{A}^m$ we obtain a matrix polynomial:

$$\mathbf{A}^N + \beta_{N-1}\mathbf{A}^{N-1} + \dots + \beta_1\mathbf{A} + \beta_0$$

Using the result $\mathbf{A} = \mathbf{U}\mathbf{\Lambda}\mathbf{U}^*$, and noting that:

$$\mathbf{A}^2 = \mathbf{U}\mathbf{\Lambda}\mathbf{U}^*\mathbf{U}\mathbf{\Lambda}\mathbf{U}^* = \mathbf{U}\mathbf{\Lambda}^2\mathbf{U}^*$$

and in general:

$$\mathbf{A}^m = \mathbf{U}\mathbf{\Lambda}^m\mathbf{U}^*$$

it follows that the matrix polynomial can be written:

$$\mathbf{U}\{\mathbf{\Lambda}^N + \beta_{N-1}\mathbf{\Lambda}^{N-1} + \beta_{N-2}\mathbf{\Lambda}^{N-2} + \dots + \beta_1\mathbf{\Lambda} + \beta_0\mathbf{I}\}\ \mathbf{U}^*$$

The diagonal matrix $\mathbf{\Lambda}$ raised to the mth power is still diagonal and contains λ_1^m, λ_2^m, etc., in its diagonal elements. Since the polynomial in λ vanishes for $\lambda = \lambda_1$, $\lambda = \lambda_2$, etc., it follows that:

$$\{\mathbf{\Lambda}^N + \beta_{N-1}\mathbf{\Lambda}^{N-1} + \beta_{N-2}\mathbf{\Lambda}^{N-2} \dots\}$$

is a null matrix with all elements zero. Hence the matrix polynomial vanishes:

$$\mathbf{A}^N + \beta_{N-1}\mathbf{A}^{N-1} + \dots + \beta_1\mathbf{A} + \beta_0 = 0 \qquad [4.8]$$

so that the matrix satisfies its own characteristic equation — this is the Cayley-Hamilton relationship.

Power methods

A result described earlier can now be explained. If an arbitrary vector is repeatedly premultiplied by the matrix $\mathbf{A}$, the result is very easily visualized if the vector is considered as a linear combination of the eigenvectors $\boldsymbol{u}_1$, $\boldsymbol{u}_2$, ..., $\boldsymbol{u}_N$.

If:

$$\boldsymbol{x} = \xi_1\boldsymbol{u}_1 + \xi_2\boldsymbol{u}_2 + \dots + \xi_N\boldsymbol{u}_N$$

then:

$$\mathbf{A}^m\boldsymbol{x} = \lambda_1^m\xi_1\boldsymbol{u}_1 + \lambda_2^m\xi_2\boldsymbol{u}_2 + \dots + \lambda_N^m\xi_N\boldsymbol{u}_N$$

Unless ξ_1 is exactly zero, then, if $|\lambda_1|$ is the largest eigenvalue, the term $\lambda_1^m \xi_1 \boldsymbol{u}_1$ eventually becomes dominant and the sequential vectors $\mathbf{A}^m \boldsymbol{x}$, $\mathbf{A}^{m+1} \boldsymbol{x}$, become simply scalar multiples of $\boldsymbol{u}_1$. The only exception occurs when the two largest eigenvalues are exactly equal. It may be remarked that the power method is a very fast and effective way for obtaining a single eigenvector but it does not give *all* the eigenvectors.

Inversion

Inversion of **A** is associated with division by eigenvalues. Writing $\boldsymbol{x}$ as was done above, one finds:

$$\mathbf{A}^{-1}\boldsymbol{x} = \lambda_1^{-1} \xi \boldsymbol{u}_1 + \lambda_2^{-1} \xi \boldsymbol{u}_2 + \ldots + \lambda_N^{-1} \xi \boldsymbol{u}_N$$

The *smallest* eigenvalue tends to become dominant in this case, and here we can identify the essential difficulty associated with matrix inversion problems: if any eigenvalue is very small, its reciprocal must be very large. If any small error creeps into a certain $\boldsymbol{x}$ which equals $\mathbf{A}\boldsymbol{v}$ — if, for example, $\boldsymbol{x}$ is derived from measurements or is computed with finite accuracy — then part of that error becomes greatly magnified. If ϵ is an error vector then the component $(\epsilon \cdot \boldsymbol{u}_N)\ \boldsymbol{u}_N$ becomes magnified by λ_N^{-1} on inversion. Even if ϵ is small, element by element, even one eigenvalue of, say, 10^{-10} will cause part of that error to be magnified ten thousand million times. We may find very huge differences in $\boldsymbol{v}$ depending on how many significant figures are retained in $\boldsymbol{x}$ when $\mathbf{A}^{-1}\ \boldsymbol{x}$ is computed, or even on minor procedural differences.

It is important to realize that this condition — to which Lanczos (1956) applied the vivid epithet "pathological" — is not caused by errors or inexact calculations. If **A** is singular, it has no inverse; if it contains a very small eigenvalue it has an inverse, but being nearly singular the inversion greatly magnifies some components. In many practical problems such instability is no better than complete singularity. One early experiment in indirect sensing with instrumental error of not more than 1% gave air temperature information by inversion with an error of several hundred degrees!

The reduction of indirect sensing problems — which usually involve integrals — to a finite system of equations generally involves some approximation. However, if the approximations involved are accurate to better than the errors of measurement, the inversion process, if it magnifies errors excessively, is not unstable solely because of the approximations. The integral equation:

$$g(y) = \int K(y, x)\, f(x)\, \mathrm{d}x$$

is almost always reduced to a system of linear equations such as:

$$g_i = \sum_j a_{ij} f(x_j)$$

and some inaccuracy results from this; but it is important to realize that the fundamental instability problem (manifested by small eigenvalues) arises from the nature of $K(y, x)$ itself. The difficulties encountered in inversion problems are very seldom caused by a loss of accuracy in going from an integral to a finite sum.

4.6. QUADRATIC FORMS, EIGENVALUES AND EIGENVECTORS

The behavior and properties of a matrix can also be examined in the light of a *scalar* function of the matrix **A**. This function is known as a "quadratic form", which is essentially the scalar (inner) product of the vectors $\boldsymbol{x}$ and $\mathbf{A}\boldsymbol{x}$. If **A** happens to be the identity matrix **I**, the quadratic form for any choice of the vector $\boldsymbol{x}$ is merely the square norm of $\boldsymbol{x}$, $|\boldsymbol{x}|^2$. In terms of the individual elements of **A** and $\boldsymbol{x}$ the associated quadratic form is $\Sigma_i \Sigma_j a_{ij}x_ix_j$; if A is a symmetric 2×2 matrix, this is:

$$a_{11}x_1^2 + 2a_1a_2x_1x_2 + a_{22}x_2^2$$

and for any $\boldsymbol{x}$ the quadratic form is a quadratic scalar function of the elements of $\boldsymbol{x}$.

Suppose now the quadratic form $\boldsymbol{x}^*\mathbf{A}\boldsymbol{x} = q$ is evaluated for all possible unitary $\boldsymbol{x}$'s (we will impose this constraint to eliminate the scalar magnitude $|\boldsymbol{x}|$ from consideration). As we go through vector space q will vary in magnitude and will display maxima and minima — there will be some vectors for which $\boldsymbol{x}^*\mathbf{A}\boldsymbol{x}$ will be lower than for all neighboring vectors, and there will be vectors for which $\boldsymbol{x}^*\mathbf{A}\boldsymbol{x}$ will be larger than for any neighboring vectors. At any of these extrema the following conditions must apply:

$$\frac{\partial q}{\partial x_1} = \frac{\partial q}{\partial x_2} = \frac{\partial q}{\partial x_3} = \ldots = \frac{\partial q}{\partial x_n} = 0$$

Now when a vector $\boldsymbol{x}$ is differentiated with respeet to one of its elements x_k the result is clearly a vector with zero elements everywhere except in the kth position, where we find a one; this particular kind of unit vector we will denote by $\boldsymbol{e}_k$.

$$\frac{\partial}{\partial x_k}\begin{bmatrix} x_1 \\ x_2 \\ \cdot \\ \cdot \\ x_k \\ x_{k+1} \\ \cdot \\ \cdot \\ x_n \end{bmatrix} = \begin{bmatrix} 0 \\ 0 \\ \cdot \\ \cdot \\ 1 \\ 0 \\ \cdot \\ \cdot \\ 0 \end{bmatrix} = \boldsymbol{e}_k$$

Returning now to the conditions for an extremum, we note that there exists the constraint $x_1^2 + x_2^2 + ..., + x_n^2 = 1$ (if there were no such constraint, the extremum found would be the trivial one $x_1 = 0, x_2 = 0$, etc., or $\boldsymbol{x} = 0$). As a result of the constraint, a small change in x_1 cannot be made independently of the values of x_2, x_3, etc. To derive the extrema under the constraint $\Sigma_j\, x_j^2 = 1$ or $\boldsymbol{x}^*\boldsymbol{x} = 1$, the most convenient procedure is the application of the method of undetermined (Lagrangian) multipliers, which in this context reduces to the finding of absolute extrema of $q - \lambda \boldsymbol{x}^*\boldsymbol{x}$, λ being an undetermined multiplier. Carrying out the succesive partial differentiations $\partial/\partial x_1$, $\partial/\partial x_2$, etc., on this quantity we obtain:

$$\frac{\partial}{\partial x_k}(q - \lambda \boldsymbol{x}^*\boldsymbol{x}) = \frac{\partial}{\partial x_k}(\boldsymbol{x}^*\mathbf{A}\boldsymbol{x} - \lambda \boldsymbol{x}^*\boldsymbol{x}) = \frac{\partial}{\partial x_k}[\boldsymbol{x}^*(\mathbf{A} - \lambda \mathbf{I})\boldsymbol{x}] = 0$$

So that:

$$\boldsymbol{e}_k^*(\mathbf{A} - \lambda \mathbf{I})\boldsymbol{x} + \boldsymbol{x}^*(\mathbf{A} - \lambda \mathbf{I})\boldsymbol{e}_k = 0$$

The second term is merely the transpose of the first when $\mathbf{A}$ is symmetric, so each must separately vanish. But $\boldsymbol{e}_k^*(\mathbf{A} - \lambda \mathbf{I})\,\boldsymbol{x}$ picks out the kth row of $(\mathbf{A} - \lambda \mathbf{I})\,\boldsymbol{x}$, and since the above equation must hold for all k, it holds for all rows of the vector and so:

$$(\mathbf{A} - \lambda \mathbf{I})\boldsymbol{x} = 0$$

or:

$$\mathbf{A}\boldsymbol{x} = \lambda \boldsymbol{x}$$

So an extremum of the quadratic form occurs when the vector $\boldsymbol{x}$ is an eigenvector of $\mathbf{A}$. The value attained by the quadratic form is then:

$$\boldsymbol{x}^*\mathbf{A}\boldsymbol{x} = \lambda \boldsymbol{x}^*\boldsymbol{x} = \lambda$$

Hence the *eigenvalue* gives the value of the quadratic form at the extremum.

This important result can be found much more directly if a rotation to principal axes is utilized. Using the relation $\mathbf{A} = \mathbf{U}\Lambda\mathbf{U}^*$ the quadratic form can be written:

$$q = \boldsymbol{x}^*\mathbf{U}\,\Lambda\,\mathbf{U}^*\boldsymbol{x}$$
$$= \boldsymbol{y}^*\,\Lambda\,\boldsymbol{y}$$

if we define a new vector $\boldsymbol{y} = \mathbf{U}^*\boldsymbol{x}$ (this will be a unit vector since, for unitary $\boldsymbol{x}$, $\boldsymbol{y}^*\boldsymbol{y} = \boldsymbol{x}^*\mathbf{U}^*\mathbf{U}\boldsymbol{x} = \boldsymbol{x}^*\mathbf{I}\boldsymbol{x} = 1$). Thus:

$$q = \lambda_1 y_1^2 + \lambda_2 y_2^2 + \lambda_3 y_3^2 + ... + \lambda_N y_N^2$$

the absolute maximum of q is λ_1 and the minimum λ_N (we envisage the λ's arranged in decreasing order of magnitude), the maximum being given by the choice $y_1 = 1$, $y_2 = y_3 = ..., = y_N = 0$, and the minimum by the choice

$y_N = 1, y_1 = y_2 = ..., = y_{N-1} = 0$. If a constraint prevents y_1 from attaining a value unity, the constrained maximum is given by assigning the rest of Σy_i^2 to y_2. If, for example, a constraint $|y_1| \leqslant \frac{1}{2}$ was applied, q would then become a maximum when $y_1 = \frac{1}{2}$ and $y_2^2 = 1 - (\frac{1}{2})^2 = \frac{3}{4}$, the remaining y's being zero. If the first m of the y's were constrained to be zero, the maximum of q would be λ_{m+1}, and it is not difficult to show that no m homogeneous constraints can lead to a maximum of q which is less than λ_{m+1}. This property is known as a minimum-maximum property, defining the $(m + 1)$th eigenvalue as the minimum value which the maximum of q can attain under m homogeneous constraints.

Geometric surfaces associated with matrices

The quadratic form $\boldsymbol{x}^* \mathbf{A}\boldsymbol{x}$ is a scalar which takes different values (for a given A) as $\boldsymbol{x}$ takes different directions in the vector space. If we change the length of the vector $\boldsymbol{x}$ as its direction changes, we can arrange for the quadratic form to remain constant; the hypersurfaces thus generated (by the end points of $\boldsymbol{x}$) are characteristic of the matrix **A** and the shape of these surfaces, if we could draw and see them, would tell us a lot about the matrix **A**. However, the limit of such visualization is three dimensions, and even three dimensions are awkward. If **A** is two-dimensional (i.e. a 2×2 matrix), and we identify x_1 and x_2 with the usual x, y coordinates of two-dimensional (plane) geometry, the equation:

$$\boldsymbol{x}^* \mathbf{A}\boldsymbol{x} = \text{constant}$$

is equivalent to:

$$a_{11}x^2 + a_{12}xy + a_{21}xy + a_{22}y^2 = \text{constant}$$

The equation is the equation of an ellipse, a parabola or a hyperbola depending on the numerical values of the coefficients. Since a_{12} and a_{21} only appear in the sum $a_{12} + a_{21}$, any matrix with $a_{12} + a_{21} =$ constant will give the same surface, but this does not create an ambiguity when the matrix is symmetric. All of the quadratic forms which will be encountered here are symmetric and therefore the question of dual systems of surfaces, which are encountered when the form is asymmetric, will not be pursued here. For symmetric **A**, the equation just written becomes:

$$a_{11}x^2 + 2a_{12}xy + a_{22}y^2 = \text{constant}$$

and the n-dimensional equation $\boldsymbol{x}^* \mathbf{A}\boldsymbol{x} =$ constant becomes:

$$a_{11}x_1^2 + a_{22}x_2^2 + ... + 2a_{12}x_1x_2 + 2a_{13}x_1x_3 + ... = \text{constant}$$

If **A** is singular there is some $\boldsymbol{x}$ such that:

$$\boldsymbol{x}^* \mathbf{A}\boldsymbol{x} = 0$$

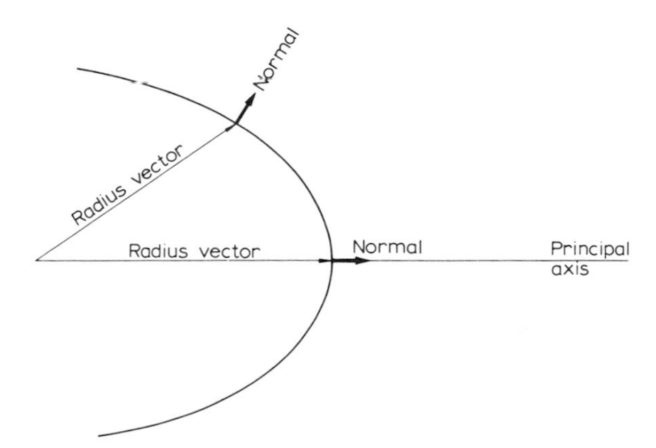

Fig. 4.2. Principal axes of quadratic surface.

so if we attempt to hold x^*Ax constant, x becomes infinitely large and the surface x^*Ax = constant recedes to infinity in that direction. When x is an eigenvector of A, x^*Ax attains an extremum and so the gradient of x^*Ax is normal to the surface, i.e. in the direction of the eigenvectors the normal and the radius vector are in the same direction and this direction defines a geometric principal axis of the quadratic surface (Fig. 4.2). Since the quadratic form attains an extremum only along the principal axes it follows that the equation x^*Ax = constant defines an *ellipsoidal* surface when A is positive-definite. If the elements of A are bounded and finite in number, the largest eigenvalue of A remains finite and so we find a surface which comes closest to the origin when the radius vector x is in the direction(s) of the eigenvector associated with the largest eigenvalue, and recedes from the origin furthest when x is in the direction(s) of the eigenvector paired with the smallest eigenvalue. The convexity of the surface x^*Ax is readily shown when A is positive definite. If x and y are two radius vectors to the surface, an arbitrary vector in the plane of x and y is obtained by a linear combination of x and y. The particular vector shown in Fig. 4.3, lying between x and y, is given by the linear combination $\beta x + (1 - \beta)y$; β must be positive and less than unity, otherwise the vector will not lie between x and y. If the end-points of all such vectors lie *inside* the surface, for all choices of x, y and for all $\beta < 1$, then convexity of the surface is established. Inserting the vector $\beta x + (1 - \beta)y$ for v in the quadratic form v^*Av, the value obtained is:

$$\begin{aligned}
&[\beta x^* + (1-\beta)y^*]A[\beta x + (1-\beta)y] \\
&\quad = \beta^2 x^*Ax + (1-\beta)^2 y^*Ay + \beta(1-\beta)(x^*Ay + y^*Ax) \\
&\quad = \beta^2 + (1-\beta)^2 + 2\beta(1-\beta)x^*Ay
\end{aligned}$$

using in the last expression the postulated symmetry of A (which makes

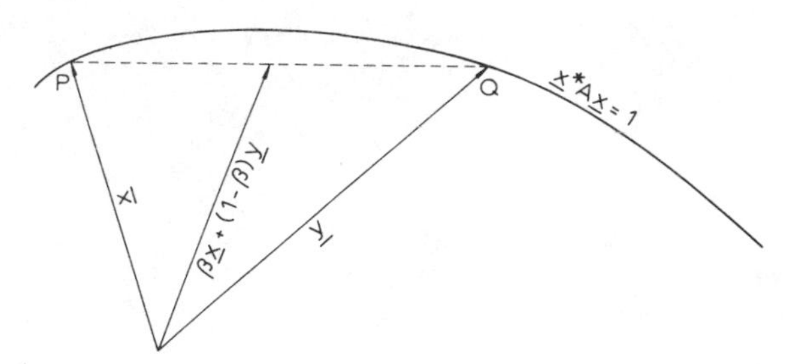

Fig. 4.3. Convex nature of quadratic surface given by $\boldsymbol{x}^*\mathbf{A}\boldsymbol{x}$ = constant when $\mathbf{A}$ is positive-definite.

$\boldsymbol{x}^*\mathbf{A}\boldsymbol{y}$ and $\boldsymbol{y}^*\mathbf{A}\boldsymbol{x}$ equal). Evidently the last expression takes the value unity at the extremities $\beta = 0$ and $\beta = 1$. Being a quadratic in β it cannot take the value 1 at a third point, so it is either >1 *everywhere* between $\beta = 0$ and $\beta = 1$ *or* it is <1 everywhere in that range. If we write p for $\boldsymbol{x}^*\mathbf{A}\boldsymbol{y}$, the quadratic in question is $\beta^2 + (1 - \beta)^2 + 2\beta(1 - \beta)\, p$ which for $\beta = \frac{1}{2}$ takes the value $\frac{1}{2} + \frac{1}{2}\, p$, which is <1 if $p < 1$. Convexity of the surface is therefore assured if $\boldsymbol{x}^*\mathbf{A}\boldsymbol{y} < 1$ when $\boldsymbol{x}^*\mathbf{A}\boldsymbol{x} = \boldsymbol{y}^*\mathbf{A}\boldsymbol{y} = 1$. That this is so can be shown in the same way that Schwarz's inequality was established: since $\mathbf{A}$ is positive definite, $\boldsymbol{x}^*\mathbf{A}\boldsymbol{x}$ is positive for every $\boldsymbol{x}$; putting $\boldsymbol{x} + \xi\boldsymbol{y}$ for $\boldsymbol{x}$, it follows that:

$$(\boldsymbol{x}^* + \xi\boldsymbol{y}^*)\,\mathbf{A}(\boldsymbol{x} + \xi\boldsymbol{y})$$

is everywhere positive; but this, a quadratic in ξ, therefore has no real roots, hence the discriminant of the quadratic must be negative, i.e. the product of the coefficients of 1 and ξ^2 must exceed one-fourth the square of the coefficient of ξ, or:

$$(\boldsymbol{x}^*\mathbf{A}\boldsymbol{x})\,(\boldsymbol{y}^*\mathbf{A}\boldsymbol{y}) > (\boldsymbol{x}^*\mathbf{A}\boldsymbol{y})^2$$

a result which may be regarded as a generalization of Schwarz's inequality. Applying this result we find that the convexity of the surface is established, since if $\boldsymbol{x}$ and $\boldsymbol{y}$ are radius vectors to the surface $\boldsymbol{x}^*\mathbf{A}\boldsymbol{x} = 1$, the quadratic form $\boldsymbol{v}^*\mathbf{A}\boldsymbol{v}$ takes a smaller value for all radius vectors with end-points lying on the line connecting the end-points of $\boldsymbol{x}$ and $\boldsymbol{y}$.

We conclude from this that for positive definite matrices the associated surfaces are hyperellipsoids, the principal axes of which coincide with the eigenvectors $\boldsymbol{u}_1, \boldsymbol{u}_2, \ldots, \boldsymbol{u}_n$ and the length of the axes of which are inversely proportional to the eigenvalues. Because of the inverse proportionality the eigenvalues were sometimes defined in an inverse way to the definition adopted here — in older literature especially the defining equation may be encountered in the form:

$$\boldsymbol{u} = \lambda\mathbf{A}\boldsymbol{u}$$

which of course leads to eigenvalues which are the reciprocals of the values given by the equation:

$$\mathrm{A}u = \lambda u \, .$$

BIBLIOGRAPHY

Bellman, R., 1970. *Introduction to Matrix Analysis.* McGraw-Hill, New York, N.Y., 403 pp.

Courant, R. and Hilbert, D., 1953. *Methods of Mathematical Physics.* Interscience, New York, N.Y., 561 pp.

Goldstine, H.H., Murray, F.J. and von Neumann, J., 1959. The Jacobi method for real symmetric matrices. *J. Assoc. Comp. Mach.*, 6: 59—66.

Goursat, E., 1933. *Course d'Analyse Mathematique.* Bautheir Villars, Paris, Vol. 2, 690 pp.

Lanczos, C., 1956. *Applied Analysis.* Prentice-Hall, Englewood Cliffs, N.J., 539 pp.

Thrall, R.M. and Tornheim, L., 1957. *Vector Spaces and Matrices.* Dover, New York, N.Y., 318 pp.

Wilkinson, J.H., 1965. *The Algebraic Eigenvalue Problem.* Clarendon Press, Oxford, 662 pp.

Zelinsky, D., 1968. *A First Course in Linear Algebra.* Academic Press, New York, N.Y., 266 pp.

CHAPTER 5

Algebraic and geometric aspects of functions and function space

The concept of length, projection, orthogonality and so on has proved to be very useful in describing and visualizing algebraic relations and it is a small and natural step to extend these concepts also into the domain of real function theory. It is, for example, useful and informative to view the expansion of an arbitrary or unknown function in a finite or infinite series as a question of determining "projections" of a function along other "directions" in function space. In connection with inversion problems this is particularly relevant since the integral equation:

$$g(y) = \int_a^b K(y, x)\, f(x)\, \mathrm{d}x$$

so often encountered in inversion problems, can be viewed as a prescription for the projection of $f(x)$ on to the directions given by $K(y, x)$, viewed as a function of x in which y is a parameter which can be varied to obtain different functions K.

In numerical work on a computer, or for that matter in graphical work, tabulation, etc., a function is always described by a collection of numbers, in other words, a vector. In many ways a function can be regarded as the limit of a sequence of vectors as the dimension of the vector goes to infinity, and in practice we tend almost always to represent a function by a vector of finite dimension; only when the function is described by an analytic formula such as $\mathrm{e}^{-\alpha x}$ can it be considered as a vector of infinite dimension (for, in principle at least, we can evaluate $\mathrm{e}^{-\alpha x}$ for any and all values of x).

In the previous chapters the dot, scalar or inner product of two vectors played a central role in relation to the important concepts of norm, length, orthogonality, linear transformations, rotations and matrix products. In forming this product the elements of the vectors are multiplied in pairs and the products added. If we increase the dimension N of the vectors without limit, this product becomes infinite, but if we include a factor Δx_N which decreases as N^{-1} this infinite growth can be prevented. This of course is simply integration, which by a familiar elementary definition is given for a function $f(x)$ being integrated from a to b by:

$$\int_a^b f(x)\, \mathrm{d}x = \lim_{N \to \infty} \sum_{k=1}^{N} f(x_k)\, \frac{b-a}{N} \qquad \left[x_k = a + (k - \tfrac{1}{2})\, \frac{b-a}{N}\right]$$

therefore, apart from a multiplicative factor, the integral of the product of two functions $f(x)$ and $g(x)$, $\int_a^b f(x)\,g(x)\,dx$ is the limit of the sequence of dot products:

$$\begin{bmatrix} f(a+\frac{1}{2}\Delta x_N) \\ f(a+\frac{3}{2}\Delta x_N) \\ \cdot \\ \cdot \\ f(b-\frac{1}{2}\Delta x_N) \end{bmatrix} \cdot \begin{bmatrix} g(a+\frac{1}{2}\Delta x_N) \\ g(a+\frac{3}{2}\Delta x_N) \\ \cdot \\ \cdot \\ g(b-\frac{1}{2}\Delta x_N) \end{bmatrix}$$

writing Δx_N for $(b-a)/N$.

It is therefore quite natural to pass over from vector spaces, spanned by all three-dimensional vectors (ordinary geometric space), by all N-dimensioned vectors (N-dimensional vector space), to function spaces, spanned by (i.e. comprising) all functions of some prescribed character on some prescribed interval of an independent variable x — a variable corresponding and analogous to the index n of the vector element v_n, not in any way similar to the x-coordinate in geometric space, which is the first (as usually written) of the elements of the three-dimensional geometric vector (x, y, z). If x, y, z were relabeled x_1, x_2, x_3 the subscripts 1, 2, 3 would be analogous to the variable x of $f(x)$: in other words, the latter corresponds to position within the N-dimensioned vector when N goes to infinity.

The manifold or set of functions contained in a prescribed function space may be, for example, all everywhere differentiable functions on the interval (0, 1), all functions of bounded variation on the interval (−1, 1) and so on. Thus there are many possible function spaces and some function spaces may include other, more restrictive, spaces. The function space spanned by all piecewise continuous functions on the interval (0, 1), for example, contains within it the space spanned by all continuous functions on the interval (0, 1).

In most, if not all, indirect sensing and inversion problems the interval of integration is finite. We will therefore not discuss in any detail the special questions posed by function spaces on infinite intervals of the dependent variable.

5.1. ORTHOGONALITY, NORMS AND LENGTH

The definition of the dot product or inner product of two functions has already been anticipated:

$$(f, g) = f \cdot g = \int_a^b f(x)g(x)\mathrm{d}x \qquad [5.1]$$

From this, of course, the definition of orthogonality follows immediately; if:

$$\int_a^b u(x)w(x)\mathrm{d}x = 0 \qquad [5.2]$$

then $u(x)$ and $w(x)$ are orthogonal on the interval (a, b). Similarly, (length)^2 or norm of $f(x)$, $|f(x)|$ or simply $|f|$, is given by:

$$|f(x)|^2 = \int_a^b f^2(x)\mathrm{d}x \qquad [5.3]$$

Just as with vectors, the Schwarz inequality is important in developing properties of functions and function spaces. It is obtained most readily by noting that if $b > a$, then:

$$\int_a^b [f(x) + \xi g(x)]^2 \mathrm{d}x$$

is necessarily non-negative for real functions. Thus:

$$\xi^2 \int_a^b g^2(x)\mathrm{d}x + 2\xi \int_a^b f(x)g(x)\mathrm{d}x + \int_a^b f^2(x)\mathrm{d}x \geqslant 0$$

The equality is a quadratic equation in ξ, of the form $\alpha\xi^2 + \beta\xi + \gamma = 0$; the theory of quadratic equations shows that quadratics have two real roots (and therefore change sign twice) if $\beta^2 > 4\alpha\gamma$. The expression above will therefore remain positive for all values of ξ if and only if:

$$\int f^2(x)\mathrm{d}x \int g^2(x)\mathrm{d}x \geqslant \left[\int f(x)g(x)\mathrm{d}x\right]^2 \qquad [5.4a]$$

This is Schwarz' inequality for functions, and can be written more concisely as:

$$(f \cdot g)^2 \leqslant |f|^2 |g|^2 \qquad [5.4b]$$

The Schmidt process (Gram-Schmidt orthogonalization)

Given a set (finite or infinite) of functions defined on an interval (a, b) in x, these functions will not in general be orthogonal, but from them we can construct linear combinations which are mutually orthogonal. The process whereby this is accomplished is again the Schmidt process. Let $f_1(x)$, $f_2(x)$, ..., $f_n(x)$ be the n functions. The process proceeds by constructing functions $\phi_m(x)$ formed as a linear combination of $f_1(x), f_2(x), ..., f_m(x)$, but not including any functions beyond $f_m(x)$, each such linear combination being orthogonal to all preceding ϕ's, $\phi_1(x)$, $\phi_2(x)$, ..., $\phi_{m-1}(x)$. During the Schmidt process it is necessary to normalize each function $\phi_m(x)$ so that its square integral over (a, b) is unity; the so normalized $\phi_m(x)$ will be written $\hat{\phi}_m(x)$, i.e. $\int [\hat{\phi}_m(x)]^2 \,dx = 1$ (the limits in the integral will not be written explicitly in the following discussion since they are always the same).

The general, mth, step in the Schmidt process consists of obtaining from $f_1(x)$, $f_2(x)$, ..., $f_m(x)$ a linear combination $\phi_m(x)$ which is orthogonal to all preceding $\hat{\phi}_k(x)$, $k = 1, 2, ..., m - 1$. That is, it is required that the coefficients $c_{m,1}, c_{m,2}, ..., c_{m,m-1}$ be chosen so that $\phi_m(x) = \sum_{k=1}^{m} c_{mk} f_k(x)$ is orthogonal with respect to $\hat{\phi}_1(x)$, $\hat{\phi}_2(x)$, ..., $\hat{\phi}_{m-1}(x)$; hence for $l < m$:

$$0 = \int \phi_m(x)\hat{\phi}_l(x)\,dx = \sum_{k=1}^{m} c_{mk} f_k(x)\hat{\phi}_l(x)\,dx$$

To ensure this result it is sufficient to subtract from $f_m(x)$ all components in it which are not orthogonal with respect to $\hat{\phi}_1(x)$, $\hat{\phi}_2(x)$, etc., just as a vector may be orthogonalized by subtracting out its components along other mutually orthogonal axes. If we form $\phi_m(x)$ as follows:

$$\phi_m(x) = f_m(x) - (f_m \cdot \hat{\phi}_1)\hat{\phi}_1(x) - (f_m \cdot \hat{\phi}_2)\hat{\phi}_2(x) - ... - (f_m \cdot \hat{\phi}_{m-1})\hat{\phi}_{m-1}(x)$$

where $(f_m \cdot \hat{\phi}_k)$ denotes the inner product $\int f_m(x)\hat{\phi}_k(x)\,dx$, then the resulting $\phi_m(x)$ is indeed orthogonal with respect to $\hat{\phi}_1(x)$, $\hat{\phi}_2(x)$, etc. For example, we have:

$$\int \phi_m(x)\phi_1(x)\,dx = \int f_m(x)\phi_1(x)\,dx - \int f_m(x)\phi_1(x)\,dx \cdot \int \hat{\phi}_1(x)\hat{\phi}_1(x)\,dx$$

which vanishes because of the normality of $\hat{\phi}_1(x)$, which implies $\int \hat{\phi}_1(x)\,\hat{\phi}_1(x)\,dx = 1$, and the mutual orthogonality of the $\hat{\phi}$'s, which causes all terms containing inner products such as $\int \hat{\phi}_1(x)\,\hat{\phi}_2(x)\,dx$ to vanish. The mth step is completed by evaluating $\int \phi_m^2(x)\,dx = |\phi_m|^2$, which allows $\phi_m(x)$ to be orthogonalized, using:

$$\hat{\phi}_m(x) = \phi_m(x)/|\phi_m|$$

Application of the Schmidt process. As an example consider the formation of polynomials in x which are orthogonal on the interval $(-1, 1)$; these polynomials are in fact the Legendre polynomials which can be evaluated also in other ways (e.g. from the Legendre differential equation or from the generating function). We have as our base functions $f_0(x)$, $f_1(x)$, etc., the powers of x: $1, x, x^2$, etc. Commencing with the subscript zero in accordance with the usual nomenclature, one has:

$$f_0(x) = 1$$

Since there are no preceding members of the sequence, $\phi_0(x)$ is simply

$f_0(x)$ or 1 and $\int_{-1}^{1} \phi_0^2(x)\, dx = 2$. Hence:

$$\hat{\phi}_0(x) = \frac{\phi_0(x)}{|\phi_0|} = \frac{1}{\sqrt{2}}$$

Turning now to $f_1(x) = x$, we have:

$$\phi_1(x) = f_1(x) - (f_1 \cdot \hat{\phi}_0)\hat{\phi}_0(x)$$

So since:

$$\int_{-1}^{1} f_1(x)\hat{\phi}_0(x)\,dx = \frac{1}{\sqrt{2}} \int_{-1}^{1} x\,dx = 0$$

$$\phi_1(x) = x, \quad \int_{-1}^{1} \phi_1^2(x)\,dx = \tfrac{2}{3}$$

and

$$\hat{\phi}_1(x) = \frac{x}{\sqrt{\frac{2}{3}}} = \sqrt{\tfrac{3}{2}}\; x$$

The next members of the sequence of base functions, $f_2(x)$, is x^2. And:

$$\phi_2(x) = x^2 - \left[\int_{-1}^{1} \frac{x^2}{\sqrt{2}}\,dx\right] \cdot \frac{1}{\sqrt{2}} - \left[\sqrt{\tfrac{3}{2}} \int_{-1}^{1} x^3\,dx\right] \sqrt{\tfrac{3}{2}}\; x$$

$$= x^2 - \tfrac{1}{3}$$

Since:

$$\int_{-1}^{1} \phi_2^2(x)\,dx = \int_{-1}^{1} (x^4 - \tfrac{2}{3}x^2 + \tfrac{1}{9})\,dx = \tfrac{2}{5} - \tfrac{4}{9} + \tfrac{2}{9} = \tfrac{8}{45}:$$

$$\hat{\phi}_2(x) = \sqrt{\tfrac{45}{8}}\,(x^2 - \tfrac{1}{3}) = \sqrt{\tfrac{5}{2}}\,(\tfrac{3}{2}x^2 - \tfrac{1}{2})$$

The Legendre polynomials as usually given are not normalized. They are:

$$P_0(x) = 1$$

$$P_1(x) = x$$

$$P_2(x) = \tfrac{3}{2}x^2 - \tfrac{1}{2}$$

$$P_3(x) = \tfrac{5}{2}x^3 - \tfrac{3}{2}x$$

The normalization factor is obtained from $\int_{-1}^{1} p_n^2(x)\, dx = 2/(2n + 1)$, which shows that the normalized functions are $\sqrt{(2n+1)/2}\; p_n(x)$, whence:

$$\hat{P}_0(x) = \sqrt{\tfrac{1}{2}}$$

$$\hat{P}_1(x) = \sqrt{\tfrac{3}{2}}\,x$$

$$\hat{P}_2(x) = \sqrt{\tfrac{5}{2}}\,(\tfrac{3}{2}x^2 - \tfrac{1}{2})$$

.
.

etc.

which agree with the orthogonal polynomials generated by the Schmidt process. The latter is very readily programmed for computer application and can provide just as easily sets of orthogonal polynomials which are orthogonal on intervals other than (—1, 1). If, for example, we follow an identical procedure but evaluate the integrals from $x = 0$ to $x = 1$ we obtain polynomials orthogonal with respect to the interval (0, 1):

$$\hat{\phi}_0(x) = 1$$

$$\hat{\phi}_1(x) = \sqrt{3}\,(2x - 1)$$

$$\hat{\phi}_2(x) = \sqrt{\tfrac{5}{29}}\,(6x^2 - 6x + 1)$$

.
.

etc.

It is important to realize that the polynomials obtained by the Schmidt process are not unique, for we arbitrarily took the base functions in the sequence 1, x, x^2, ..., and any change in this order will give a different set of polynomials. If, for example, we took x^2, x and 1 in that order the Schmidt process would proceed as follows:

(1) $f_1(x) = x^2 = \phi_1(x)$

$$\int_0^1 \phi_1^2(x)\,\mathrm{d}x = \tfrac{1}{5}; \qquad \hat{\phi}_1(x) = \sqrt{5}\,x^2$$

(2) $f_2(x) = x; \qquad (f_2 \cdot \hat{\phi}_1) = \sqrt{5}\int_0^1 x^3\,\mathrm{d}x = \dfrac{\sqrt{5}}{4}$

$$\phi_2(x) = x - \frac{\sqrt{5}}{4}\,\hat{\phi}_1(x) = x - \tfrac{5}{4}x^2$$

$$\int_0^1 [\phi_2(x)]^2\,\mathrm{d}x = \tfrac{1}{48}$$

$$\hat{\phi}_2(x) = \sqrt{48}\,(x - \tfrac{5}{4}x^2) = 4\sqrt{3}\,x - 5\sqrt{3}\,x^2$$

(3) $f_3(x) = 1; \qquad (f_3 \cdot \hat{\phi}_1) = \dfrac{\sqrt{5}}{3}; \qquad (f_3 \cdot \hat{\phi}_2) = \dfrac{\sqrt{3}}{3}$

$$\phi_3(x) = 1 - \frac{\sqrt{3}}{3}(4\sqrt{3}\,x - 5\sqrt{3}\,x^2) - \frac{\sqrt{5}}{3}\sqrt{5}\,x^2$$

$$= 1 - 4x + \tfrac{10}{3}x^2$$

$$\int [\phi_3(x)]^2 \mathrm{d}x = \tfrac{1}{9}$$

$$\hat{\phi}_3(x) = 3 - 12x + 10x^2$$

The three polynomials are mutually orthogonal and normalized but evidently are different from the polynomials obtained from the same base functions taken in the reverse order.

If a set of N functions is prescribed there is no guarantee that the set will be independent; it may happen that one function is a linear combination of some or all of the others. If the Schmidt process is applied to such a set then at some stage removal from a certain $f_k(x)$ of its non-orthogonal part will reduce it to zero, so that $\phi_k(x)$ becomes indeterminate. That particular $f_k(x)$ can simply be dropped from the set, for it is redundant; since:

$$\phi_k(x) = 0 = f_k(x) - \alpha_1\hat{\phi}_1(x) - \alpha_2\hat{\phi}_2(x) - \ldots - \alpha_{k-1}\hat{\phi}_{k-1}(x)$$

and the $\hat{\phi}$'s are linear combinations of f's of lower or equal order, it follows that $f_k(x)$ must be identically a linear combination of $f_1(x)$, $f_2(x)$, ..., $f_{k-1}(x)$. It follows that if m of a set of N functions can be dropped in this way, there are only $N - m$ independent components and the dimension of the function space spanned by all possible combinations of the original N function is $N - m$.

When following the Schmidt procedure the function $\hat{\phi}_1(x)$ was proportional to $f_1(x)$, $\hat{\phi}_2(x)$ was a linear combination of $f_1(x)$ and $f_2(x)$, $\hat{\phi}_3(x)$ a linear combination of $f_1(x)$, $f_2(x)$ and $f_3(x)$, and so on. That is, the coefficients relating the ϕ's to the f's constitute a triangular array — one non-vanishing coefficient in the first row, two in the second, etc. If the functions $f_1(x)$, $f_2(x)$, ..., $f_N(x)$ are collected into a vector $\boldsymbol{f}(x)$ and the functions $\hat{\phi}_1(x)$, $\hat{\phi}_2(x)$, ..., $\hat{\phi}_N(x)$ into a vector $\hat{\boldsymbol{\phi}}(x)$, then:

$$\hat{\boldsymbol{\phi}}(x) = \mathbf{T}\boldsymbol{f}(x)$$

where $\mathbf{T}$ is a lower triangular matrix (i.e. $t_{ij} = 0$ when $j > i$). The orthonormality of the $\hat{\phi}$'s, implying that $\int \hat{\phi}_i(x)\,\hat{\phi}_j(x)\,\mathrm{d}x = 0$ except when $i = j$, when it is unity, means that the matrix $\|\int \phi_i(x)\,\phi_j(x)\,\mathrm{d}x\|$ is the identity matrix $\mathbf{I}$. That matrix is obtained from the vector $\hat{\boldsymbol{\phi}}(x)$ by forming the "outer" product $\hat{\phi}(x)\,\hat{\phi}^*(x)$ and integrating. Hence since **T** is not a function of x:

$$\mathbf{I} = \int \hat{\phi}(x)\hat{\phi}^*(x)\,\mathrm{d}x = \mathbf{T}\,[\int f(x) f^*(x)\,\mathrm{d}x\,]\mathbf{T}^*$$

the bracketed term is the so-called covariance matrix **C** of the functions $f_1(x)$, $f_2(x)$, ..., $f_N(x)$, formed by multiplying the functions in pairs and integrating. The Schmidt process therefore represents a diagonalization of the covariance matrix **C**. Many other diagonalizations are possible and any matrix **A** which produces the result:

$$\mathbf{ACA}^* = \mathbf{I}$$

gives an orthonormal set of functions $\hat{\phi}_1$, $\hat{\phi}_2$, ... with:

$$\hat{\phi}(x) = \mathbf{A} f(x)$$

5.2. OTHER KINDS OF ORTHOGONALITY

The procedures and properties of the previous section all hinge on the definition of the inner products. The definition used is the simplest and commonest, but certainly not the only definition possible for the inner product of two functions. One can for instance define the orthogonality of two functions to be given by:

$$\int w(x) f_m(x) f_n(x)\,\mathrm{d}x = 0$$

This is *weighted orthogonality*, $w(x)$ being the weighting function. It is equivalent to ordinary, unweighted orthogonality of the functions $f_m(x)\sqrt{w(x)}$ and $f_n(x)\sqrt{w(x)}$.

Weighted orthogonality is useful to obtain orthogonal functions on the infinite intervals $(0, \infty)$ or $(-\infty, \infty)$. It is obviously not possible to find polynomials which are orthonormal on these intervals since $\int_0^\infty x^p\,\mathrm{d}x$ and $\int_{-\infty}^{\infty} x^p\,\mathrm{d}x$ diverge. But the integral $\int_0^\infty x^p\,\mathrm{e}^{-x}\,\mathrm{d}x$ exists and so does $\int_{-\infty}^{\infty} x^p\,\mathrm{e}^{-x^2}\,\mathrm{d}x$, and orthogonal polynomials can be derived by introducing e^{-x} as a weighting function in the case of the interval $(0, \infty)$ or e^{-x^2} for the interval $(-\infty, \infty)$; it is of course also possible but less useful to use e^{-x^2} for the interval $(0, \infty)$.

Laguerre polynomials

Using the weight function e^{-x}, so that orthogonality of $f_m(x)$ and $f_n(x)$ implies $\int_0^\infty \mathrm{e}^{-x} f_m(x) f_n(x)\,\mathrm{d}x = 0$, a set of polynomials can be constructed just as readily as the Legendre polynomials were constructed. These polynomials are known as Laguerre polynomials. In constructing these polynomials, terms of the form $\int_0^\infty x^p\,\mathrm{e}^{-x}\,\mathrm{d}x$ will need to be calculated. This is by definition the gamma function $\Gamma(p+1)$, and for integer p the numerical values are very simply obtained using the initial value $\int_0^\infty e^{-x}\,dx = 1$ and the recursive relationship:

$$\int_0^\infty x^p e^{-x}\,dx = p\int_0^\infty x^{p-1} e^{-x}\,dx$$

which is obtained by integration by parts. We therefore have:

$$\int_0^\infty x\,e^{-x}\,dx = 1,\ \int_0^\infty x^2 e^{-x}\,dx = 2,\ \ldots\ \int_0^\infty x^p e^{-x}\,dx = p!$$

Applying the Schmidt process we have:

(1) $f_0(x) = 1;\qquad \phi_0(x) = 1$

$$\int_0^\infty [\phi_0(x)]^2 e^{-x}\,dx = 1$$

$$\hat{\phi}_0(x) = 1$$

(2) $f_1(x) = x;\qquad \int_0^\infty f_1(x)\hat{\phi}_0(x)\,e^{-x}\,dx = \int_0^\infty x\,e^{-x}\,dx = 1$

$$\phi_1(x) = x - 1$$

$$\int_0^\infty [\phi_1(x)]^2 e^{-x}\,dx = \int_0^\infty x^2 e^{-x}\,dx - 2\int_0^\infty x\,e^{-x}\,dx + \int_0^\infty e^{-x}\,dx = 2-2+1 =$$

$$\hat{\phi}_1(x) = x - 1$$

(3) $f_2(x) = x^2;\qquad \int_0^\infty f_2(x)\hat{\phi}_0(x)\,e^{-x}\,dx = 2;$

$$\int_0^\infty f_2(x)\hat{\phi}_1(x)e^{-x}\,dx = 3!-2! = 4$$

$$\phi_2(x) = x^2 - 4x + 2$$

$$\int_0^\infty [\phi_2(x)]^2 e^{-x}\,dx = \int_0^\infty (x^4 - 8x^3 + 20x^2 - 16x + 4)e^{-x}\,dx$$

$$= 4! - 8.3! + 20.2! - 16 + 4 = 4$$

$$\hat{\phi}_2(x) = \tfrac{1}{2}(x^2 - 4x + 2)$$

These are the first few normalized Laguerre polynomials.

In the case of the interval $(-\infty, \infty)$ and a weighting function e^{-x^2}, the orthonormal polynomials are the *Hermite* polynomials. They can be derived just as the Laguerre polynomials were using $\int_{-\infty}^{\infty} x^m\ e^{-x^2}\ dx = \frac{1}{2}(m-1) \int_{-\infty}^{\infty} x^{m-2}\ e^{-x^2}\ dx$ and starting values $\int_{-\infty}^{\infty} e^{-x^2}\ dx = \sqrt{\pi}$ and $\int_{-\infty}^{\infty} e^{-x^2}\ dx = 0$. The first few normalized *Hermite* polynomials are:

$$\hat{\phi}_0(x) = \pi^{-1/4}$$

$$\hat{\phi}_1(x) = \sqrt{2}\,\pi^{-1/4}x$$

$$\hat{\phi}_2(x) = \sqrt{\tfrac{4}{5}}\,\pi^{-1/4}(x^2 - \tfrac{1}{2})$$

Summation orthogonality

Another kind of orthogonality exists when the sum $\Sigma_k\ f_m(x_k)\ f_n(x_k)$ vanishes. The x_k may be evenly spaced or unevenly spaced so a very great range of possibilities exists for functions to display this kind of orthogonality. Many of the transcendental functions of mathematics possess both this kind of orthogonality and orthogonality (usually weighted orthogonality) with respect to integration. Bessel functions for instance satisfy the weighted orthogonality

$$\int_0^1 x\,J(\alpha_m x)\,J(\alpha_n x)\,dx = 0 \qquad (m \neq n)$$

(α_m, α_n being roots of $J(x)$) and similar relationships can be derived for a wide range of functions. The existence of a linear second order differential equation defining the function usually implies that two functions satisfying the same boundary conditions must be orthogonal.

The trigonometric functions possess orthogonality with respect to both integration and summation, properties which have widespread application in Fourier analysis, trigonometric interpolation and spectral analysis. For integers m and n we have for $m \neq n$:

$$\int_{-\pi}^{\pi} \cos mx \cos nx\ dx = 0 \text{ and } \int_{-\pi}^{\pi} \sin mx \sin nx\ dx = 0$$

while for any m and n other than $m = n = 0$ we have:

$$\int_{-\pi}^{\pi} \cos mx \sin nx\ dx = 0$$

These orthogonalities can be proved in several ways. Taking by way of illustration the first orthogonality given above (that of cos mx and cos nx), this may readily be derived by writing:

$\cos mx \cos nx = \frac{1}{2}\cos(m+n)x + \frac{1}{2}\cos(m-n)x$

Now:

$$\int_{-\pi}^{\pi} \cos(m+n)x\,\mathrm{d}x = \left|\frac{\sin(m+n)x}{m+n}\right|_{-\pi}^{\pi}$$

which vanishes unless $m = -n$; similarly $\int_{-\pi}^{\pi} \cos(m-n)x\,\mathrm{d}x$ vanishes unless $m = n$. Thus unless $|m| = |n|$, cos mx and cos nx are orthogonal. When $m = n$ we obtain the normalizing factors:

$$\int_{-\pi}^{\pi} \cos mx \cos mx\,\mathrm{d}x = \frac{1}{2}\int_{-\pi}^{\pi} (1 + \cos 2mx)\,\mathrm{d}x = \pi$$

except when $m = 0$, in which case the integral is 2π.
To prove the summation orthogonality of the trigonometric functions, it is simplest to look at the product of the functions e^{imx} and e^{inx} summed over the $2N + 1$ points $= -N\xi, (-N+1)\xi, \ldots, (N-1)\xi, N\xi$. This is not the most general situation possible, but the extension of the result is routine if somewhat more tedious. We have:

$$\sum_{k=-N}^{N} e^{i(m+n)k\xi} = \sum_{k=-N}^{N} [e^{i(m+n)\xi}]^k$$

a geometric series which has the sum:

$$\frac{e^{i(m+n)\xi(N+1)} - e^{-i(m+n)\xi N}}{e^{i(m+n)\xi} - 1}$$

or:

$$\frac{e^{i(m+n)\xi(N+1/2)} - e^{-i(m+n)\xi(N+1/2)}}{e^{i(m+n)\xi/2} - e^{-i(m+n)\xi/2}} = \frac{\sin[(m+n)\xi(N+\frac{1}{2})]}{\sin[(m+n)\xi/2]}$$

So if $m + n \neq 0$, this sum is zero provided $\sin[(m+n)\xi(N+\frac{1}{2})]$ vanishes, i.e. if $(m+n)\xi(N+\frac{1}{2}) = \pi$ (or a multiple thereof). If $\xi(N+\frac{1}{2}) = \pi$, the sum will vanish for all integer and non-zero values of $m + n$. With this choice for ξ we have $\xi = 2\pi/(2N+1)$, so that the tabular x_k values are spaced evenly between $\pm\pi/(1+\frac{1}{2}N)$. (If $N = 5$, for example, the 11 tabular points are

$x = -\frac{10}{11}\pi, -\frac{8}{11}\pi, -\frac{6}{11}\pi, -\frac{4}{11}\pi, -\frac{2}{11}\pi, 0, \frac{2}{11}\pi, \frac{4}{11}\pi, \frac{6}{11}\pi, \frac{8}{11}\pi, \frac{10}{11}\pi$.)

Over these tabular points in x, the sum $\Sigma_k\, e^{imx}\, e^{inx}$ vanishes for

$(m + n) \neq 0$. The real part of this:

$$\sum_k (\cos mx_k \cos nx_k - \sin mx_k \sin nx_k)$$

must vanish, as also the imaginary part:

$$\sum_k (\cos mx_k \sin nx_k + \sin mx_k \cos nx_k)$$

but this result must also hold when n is replaced by $-n$ (unless $m - n = 0$), a replacement which changes $\sin nx_k$ in sign but leaves $\cos nx_k$ unaltered; hence if $|m|$ and $|n|$ are unequal:

$$\left.\begin{aligned} \sum_k \cos mx_k \cos nx_k &= 0 \\ \sum_k \sin mx_k \sin nx_k &= 0 \\ \sum_k \cos mx_k \sin nx_k &= 0 \end{aligned}\right\}$$

If $m + n = 0$ the sum of the geometric series becomes zero in numerator and denominator, but the ratio is easily seen to be finite and equal to $2N + 1$. Hence, for $m + n = 0$.

$$\sum_k \cos^2 mx_k + \sum_k \sin^2 mx_k = 2N + 1$$

$$\sum_k \cos^2 mx_k - \sum_k \sin^2 mx_k = 0$$

So:

$$\left.\begin{aligned} \sum_k \cos^2 mx_k &= \frac{2N+1}{2} \\ \sum_k \sin^2 mx_k &= \frac{2N+1}{2} \end{aligned}\right\}$$

while

$$\sum_k \cos mx_k \sin mx_k = 0$$

Closely allied to the trigonometric function are the Chebychev polynomials. These are simply the trigonometric-functions $\cos n\theta$, $\sin n\theta$ with $\cos\theta$ taken as the independent variable x. The trigonometric recurrence relationship (often called Chebychev's formula):

$$\cos(n+1)\theta = 2\cos\theta\cos n\theta - \cos(n-1)\theta$$

shows that $\cos n\theta$ for any integer n can be expressed as a polynomial in $\cos\theta$. The Chebychev polynomial $T_n(x)$, defined as $2^{-n+1}\cos n(\cos^{-1} x)$ is therefore a polynomial of order n in x. The interval $-\pi \leqslant \theta \leqslant \pi$ corresponds to $-1 \leqslant x \leqslant 1$.

The summation orthogonality of the trigonometric functions was displayed over evenly spaced intervals of the (angular) argument from $-\pi$ to π; in terms of x these become unevenly spaced, being closest near $x = \pm 1$ and more sparse around $x = 0$, so the summation orthogonality of the Chebychev polynomials is displayed over a non-uniform array of points. They have the following orthogonality properties:

$$\sum_{k=-N}^{+N} T_m(x_k)\,T_n(x_k) = 0 \qquad \left(m \neq n,\ x_k = \cos^{-1}\frac{2k}{2N+1}\right)$$

$$\int_{-1}^{1} T_m(x)\,T_n(x)\,(1-x^2)^{-1/2}\,dx = 0$$

There is of course no fundamental distinction between the Chebychev polynomials and the trigonometric functions (we have only mentioned one kind of Chebychev polynomials, that corresponding to the cosine, but others also obviously follow in a similar way). It is a matter largely of choice whether one uses an interval $(-\pi, \pi)$ and trigonometric functions or an interval $(-1, 1)$ and Chebychev polynomials.

5.3. APPROXIMATION BY SUMS OF FUNCTIONS

A function $f(x)$ may be prescribed in a number of ways, some explicit, others implicit. For a finite interval a function can be graphed or tabulated — explicit up to a point, but implying that the behavior can be prescribed in between the tabled values, or in the case of a graph that there is no fine structure beyond the resolution of the graph. Functions may be defined implicitly by differential equations, by difference equations, by integrals and by integral equations. In many instances several equivalent definitions exist and it is not always obvious that the definitions are equivalent — Bessel functions, for example, may be defined by a differential equation and also by Sommerfeld's integral.

It is often useful and sometimes necessary to represent or approximate functions by means of sums of other functions. Many differential equations are solved by substituting a power series $A_0 + A_1x + A_2x^2 + \ldots$ into the differential equation and thereby finding a recursive relationship between the A's. Other differential equations are not amenable to power series substitution but are simplified by use of a Fourier series $A_0 + A_1 \cos x + A_2 \cos 2x + \ldots$ in the equation defining the function.

Some equations, for example Legendre's differential equation:

$$(x^2 - 1)\frac{d^2y}{dx^2} + 2x\frac{dy}{dx} - n(n+1)y = 0$$

are solved by a finite sum of powers of x, i.e. a polynomial. Another possibility is that a solution although strictly an infinite series can be approximated closely by a small number of leading terms of the infinite series, which therefore provides a polynomial approximation. Classical mathematical analysis is based to a large extent on the representation of functions by a power series. Such a power series is an exact representation if taken to an infinite number of terms, and at $x = 0$ the function $f(x)$ must equal the first term A_0 of the power series. As we go away from the origin, higher and higher order terms must be included to achieve some prescribed accuracy. It should be realized clearly, however, that truncation of the power series after a finite number of terms is only one way of obtaining a polynomial approximation, and it makes no effort to "spread" the error over a finite interval in x, sacrificing some accuracy at $x = 0$ to improve the accuracy elsewhere. Lanczos (1956) shows, for example, that the function x^6 (which is essentially a power series with only one non-vanishing term) can be approximated very closely throughout the interval $0 \leqslant x \leqslant 1$ by the fifth-degree polynomial:

$$3x^5 - 3.375x^4 + 1.75x^3 - 0.410156x^2 + 0.035156x - 0.000488$$

which does not pass through the origin and is quite different to a truncated power series, which will give zero when truncated anywhere before the sixth-power term.

From the point of view of function space the existence of approximations such as the fifth-order approximation for x^6 arises from the skewness (i.e. non-orthogonality and mutual interdependence) of the base functions 1, x, x^2, x^3, x^4, x^5, x^6. In an orthogonal system of axes prescribing the value of one coordinate does not give any information whatever about the value of the other coordinates but if the axes are highly skewed this is no longer true.

In obtaining approximations to functions, orthogonal functions possess the same advantages that are possessed by orthogonal coordinate axes in geometry, so are desirable for that reason alone. As it happens they also simplify the process of determining the coefficients in the approximation. Consider the representation of a function $f(x)$ by a sum of other functions $u_1(x)$, $u_2(x)$, etc.

$$f(x) \cong \xi_1 u_1(x) + \xi_2 u_2(x) + \ldots + \xi_N u_N(x) \qquad [5.5]$$

Assume first that the u's are orthonormal — if not, the Schmidt process enables the production of linear combinations of the u's which are orthonormal. If the equality sign applies in equation [5.5] then taking the inner product of both sides with $u_k(x)$ — i.e. multiplying by $u_k(x)$ and integrating over x, one finds:

$$\int f(x) u_k(x) \mathrm{d}x = \xi_k \qquad [5.6]$$

so the expansion coefficients are very easily found when the base functions $u_1(x)$, $u_2(x)$, etc., are orthonormal. It is not difficult to show that if the identity is not exact, the coefficients given by [5.6] are still optimal, in that the ξ's so chosen given a minimum value for the "unexplained" residual $f(x) - \Sigma \xi_k u_k(x)$. For if we denote this residual by $r(x)$ we have:

$$|r|^2 = \int [f(x) - \sum_k \xi_k u_k(x)]^2 \mathrm{d}x$$

This is made a minimum by choosing the ξ's so as to make $(\partial/\partial \xi_1)|r|^2 = 0$, $(\partial/\partial \xi_2)|r|^2 = 0$, and so on. The result is the Gauss or "normal" equations:

$$\left.\begin{aligned}
&\xi_1 \int u_1^2(x)\mathrm{d}x + \xi_2 \int u_1(x)u_2(x)\mathrm{d}x + \ldots + \xi_N \int u_1(x)u_N(x)\mathrm{d}x = \int u_1(x) f(x)\mathrm{d}x \\
&\xi_1 \int u_1(x)u_2(x) + \xi_2 \int u_2^2(x)\mathrm{d}x + \ldots + \xi_N \int u_2(x)u_N(x)\mathrm{d}x = \int u_2(x) f(x)\mathrm{d}x \\
&\cdot \\
&\cdot \\
&\xi_1 \int u_1(x)u_N(x)\mathrm{d}x + \xi_2 \int u_2(x)u_N(x)\mathrm{d}x + \ldots + \xi_N \int u_N^2(x)\mathrm{d}x = \int u_N(x) f(x)\mathrm{d}x
\end{aligned}\right\} \qquad [5.7]$$

In these equations we have not yet made use of the orthogonality of the u's, so equations [5.7] are applicable to find optimal (so-called "least-squares") approximations whatever the functions $u_k(x)$ may be. But if the u's are an orthonormal set, the off-diagonal elements in the Gauss equations all vanish and the equations become:

$$\xi_1 = \int f(x) u_1(x) \mathrm{d}x$$

$$\xi_2 = \int f(x) u_2(x) \mathrm{d}x$$

$$\cdot$$

$$\cdot$$

$$\xi_N = \int f(x) u_N(x) \mathrm{d}x$$

so the proposition that the coefficients given by [5.6] were optimal has been established.

Orthogonal functions possess a further advantage in connection with the magnitudes of the coefficients in equation [5.5]. If the base functions are not orthogonal there is no a priori limit to the magnitude of the ξ_k — and it is a familiar experience in series expansions to find very large coefficients alternating in sign even though the function being approximated may nowhere exceed unity. The use of orthogonal functions for the u's gives:

$$\int [f(x)]^2 \mathrm{d}x = \int [\xi_1 u_1(x) + \xi_2 u_2(x) + \ldots + \xi_N u_N(x) + r(x)]^2 \mathrm{d}x$$

$$= \xi_1^2 + \xi_2^2 + \ldots + \xi_N^2 + |r|^2$$

so the sums of squares of coefficients is bounded.

Highly skewed base functions

Fig. 5.1 depicts a two-dimensional situation where a point P is located with respect to orthogonal axes OX, OY in (a) and highly skewed axes OX, OY' in (b). There is formally no difference between prescribing the coordinates of P as x, y (with respect to the axes OX, OY) or x', y' (with respect to the axes OX, OY') — either give P uniquely. Suppose now, however, that there is some uncertainty associated with our knowledge of the coordinates, as indicated by the strips in Fig. 5.1a, b. In each case P is located somewhere within the black area, and it is obvious that in the skew frame of reference the vertical position of P is subject to considerable uncertainty. Furthermore as the axes are made more and more skew, this uncertainty increases more and more — and it so happens that there is no appreciable improvement in the determination of horizontal position.

This very simple geometric fact is at the root of most problems associated

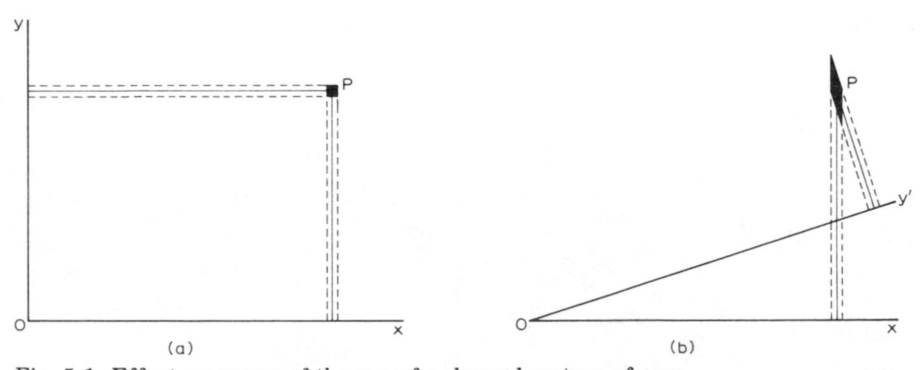

Fig. 5.1. Effect on errors of the use of a skewed system of axes.

with inversions and solution of linear Fredholm integral equations of the first kind, such as that written down at the start of this chapter. The corresponding situation in function space arises if two base functions are unitary but almost parallel, in the sense that if the second, $v_2(x)$, is written $v_2(x) = \alpha_1 v_1(x) + \alpha_2 \phi_2(x)$, $\phi_2(x)$ being orthogonal with respect to $v_1(x)$, then $|\alpha_2| << |\alpha_1|$. Suppose some function $f(x)$ is such that:

$$f(x) = a_1 v_1(x) + a_2 \phi_2(x)$$

Let $f(x)$ be measured by obtaining the inner products of $f(x)$ with the skewed base function pair $v_1(x)$ and $v_2(x)$, i.e. the projection of $\boldsymbol{f}$ on the bases $\boldsymbol{v}_1$ and $\boldsymbol{v}_2$. If there is any error in the measurements (and a real measurement will always be subject to error), then the result of the measurements will be:

$$\int v_1(x) f(x) \mathrm{d}x = b_1 + \epsilon_1$$

$$\int v_2(x) f(x) \mathrm{d}x = b_2 + \epsilon_2$$

ϵ_1 and ϵ_2 being errors. To obtain a_1 and a_2 we must have:

$$a_1 = \int f(x) v_1(x) \mathrm{d}x = b_1 + \epsilon_1$$

$$a_2 = \int f(x) \phi_2(x) \mathrm{d}x = \frac{1}{\alpha_2} (b_2 + \epsilon_2 - \alpha_1 b_1 - \alpha_1 \epsilon_1)$$

When $|\alpha_2| << |\alpha_1|$, the error in a_2 is greatly amplified. For an extreme numerical example, take $\alpha_2 = 0.01$, $\alpha_1 = 0.99995$ (to eight-figure accuracy); if the correct values of a_1 and a_2 are both unity, b_1 and b_2 are evidently 1 and 1.00995. Inserting these values in the above expressions for a_1 and a_2 we find (for no error):

$$a_1 = 1$$

$$a_2 = 100(1.00995 - 0.99995) - 1$$

But if $\epsilon_1 = -0.01$ and $\epsilon_2 = 0.01$, we find that the error in a_2, i.e. $\alpha_2^{-1}(\epsilon_2 - \alpha_1 \epsilon_1)$ is 1.99995, or almost twice as large as a_2 itself. This error amplification is the exact analogue of the geometric position error amplification shown diagramatically in Fig. 5.1.

The theory of algebraic systems of equations shows that a necessary and sufficient condition for the singularity of the Gauss equations [5.7] is the vanishing of the determinant:

$$\begin{vmatrix} \int u_1^2(x)\mathrm{d}x & \int u_1(x)u_2(x)\mathrm{d}x & \dots \\ \cdot & \cdot & \dots \\ \cdot & \cdot & \dots \\ & & \int u_N^2(x)\mathrm{d}x \end{vmatrix} = \det(\| \int u_i(x)u_j(x)\mathrm{d}x \|)$$

The geometric interpretation of this determinant is the volume of an N-dimensional parallelopiped. In two-dimensional geometric space two vectors $\boldsymbol{u}_1, \boldsymbol{u}_2$ subtend a parallelogram of area $u_1 u_2 \sin\theta$; the determinant of inner products gives:

$$\begin{vmatrix} \boldsymbol{u}_1 \cdot \boldsymbol{u}_1 & \boldsymbol{u}_1 \cdot \boldsymbol{u}_2 \\ \boldsymbol{u}_2 \cdot \boldsymbol{u}_1 & \boldsymbol{u}_2 \cdot \boldsymbol{u}_2 \end{vmatrix} = u_1^2 u_2^2 (1 - \cos^2\theta) = [\text{area}]^2$$

and in function space of N dimensions the determinant (known as the Gram determinant) essentially measures N-dimensional $(volume)^2$. If any base function is a linear combination of the others, the determinant can be transformed into one with two equal rows (or columns), which the theory of determinants shows to vanish. If all the base functions are orthonormal the determinant contains ones along the diagonal and zero everywhere else, hence is unity. The degree to which the determinant deviates from unity indicates the degree of skewness of the set of base functions.

5.4. INTEGRAL EQUATIONS

The integral equations of indirect sensing problems with kernels $K(y, x)$ can be looked on from the function space point of view as providing distinct (but perhaps highly skewed) base functions as y is varied. The measurement of $\int K(y, x)\ f(x)\ \mathrm{d}x$ provide the "projections" of the unknown $f(x)$ on the system of base functions consisting of $K(y, x)$ at different values of y. There is, however, no assurance that the dimension of the function space continues to increase as y varies along the real axis from zero to one or even from zero to infinity or minus infinity to plus infinity. The point may be put in another way: the function space spanned by all admissible $f(x)$ is mapped by the transformation $g(y) = \int K(y, x)\ f(x)\ \mathrm{d}x$ onto a "smaller", more restrictive function space. The integral operator for example transforms piecewise continuous functions $f(x)$ into continuous functions when K is a continuous function of y. The kernels of real problems change relatively smoothly and slowly with y and if we have a measurement at one y it will be necessary to change y by a finite amount before appreciable change in $g(y)$ can occur. Thus since the range of y is necessarily restricted, we really have only a finite number of measurements which can be made and the problem becomes that of determining $f(x)$, or some useful approximation or estimate for $f(x)$, from N measurements which give the "projections" $\int K_i(x)\, f(x)\, dx$ for $i = 1, 2, ..., N$. If the K's were orthonormal the solution process would be straightforward, and the inference of $f(x)$ in such a situation is not normally considered an "inversion" problem. Examples of orthogonal $K_i(x)$ are provided, for example, in interferometers or interferometer-spectrometers where the Fourier transform $\int_{-\infty}^{\infty} e^{iyx}\, f(x)\, dx$ of a desired spectral function $f(x)$ is measured — at least over a finite interval in y — and the inversion is a question of applying a further Fourier transform to the measured $g(y)$. The errors in the solution are similar in magnitude to those in the measurement. The transformation $g \to f$ is unitary, being $\boldsymbol{f} = \mathbf{W}\boldsymbol{g}$ in vector notation, with $\mathbf{W}^*\mathbf{W} = \mathbf{I}$, so that the error in $\boldsymbol{f}$ is related to that (ϵ) in $\boldsymbol{g}$ by $|\delta \boldsymbol{f}|^2 = \epsilon^* \mathbf{W}^* \mathbf{W} \epsilon = |\epsilon|^2$.

Indirect sensing measurements usually involve kernels which are inherently exponential in character, involving as they do some kind of decay (radiation decaying in intensity as it travels through an absorbing medium, particle concentrations decaying during passage through a filter) and are closer in character to the Laplace transform $g(y) = \int_0^{\infty} e^{-yx}\, f(x)\, dx$ than to the Fourier transform. Although the transforms are formally very similar, y ranges in one case along the real axis and in the other along the imaginary axis; in the former case (the Laplace transform) there are no roots of e^{-yx} along the path of integration. The kernels $e^{-y_1 x}$, $e^{-y_2 x}$, etc., being positive throughout cannot therefore be orthogonal whatever the values of the (real) parameter y_i and they necessarily provide a skewed non-orthogonal system of base functions in function space.

Since the Schmidt process provides a means of orthogonalizing arbitrary functions, one might well ask — why not orthogonalize the $K_i(x)$ in this way and retrieve all the virtues of orthogonal kernels? This certainly can be done, but when it is done the coefficients of the original kernels prove to be very large, that is, the orthogonal $\phi_i(x)$ are given by expressions $\Sigma c_{ij} K_j(x)$ in which

TABLE 5.1
Coefficients of the functions $x \exp(-y_i x)$ in orthonormal functions of the Schmidt process

i	y_i^{-1}	Coefficients					
1	0.05	179					
2	0.10	−275	116				
3	0.15	371	−374	153			
4	0.20	−468	824	−773	274		
5	0.25	568	−1510	2400	−1900	584	
6	0.30	−670	2500	−5870	7680	−5160	1390

the individual c_{ij} can be very large. When applied to measured g_j which contain errors ϵ_j the error contribution can become intolerably large. To illustrate this Table 5.1 shows the values of c_{ij} obtained using the kernels $x\ e^{-y_i x}$, with y_i^{-1} taking the values 0.05, 0.1, 0.15, ..., 0.95 (the kernels therefore attain their single maxima at these values of the independent variable).

Quadrature

At the beginning of this chapter it was pointed out that functions could be viewed as vectors carried to a limit of infinite dimensions and integral operators as a limit of matrix operators. The operation of numerical quadrature bridges the gap between vectors and functions, matrices and integral transforms, in the reverse direction, for it represents (approximately, but in many instances quite accurately) an integral as a sum of a finite number of terms. Indeed if an integral cannot be evaluated analytically, that is the only way a numerical value can be obtained for it; one writes:

$$\int_a^b f(x)\mathrm{d}x \cong \sum_{j=1}^{N} w_j f(x_j) \qquad [5.8]$$

selecting tabular x_1, x_2, ..., x_N and quadrature coefficients w_j so as to make the approximation as close as possible. The x_j's may be equidistant, or the spacing between them may vary and there are as a result of the many possible combinations many quadrature formulae. With the advent of high-speed computers it is now less relevant to seek out optimum quadrature formulae and quite often it is better to use a simple quadrature formula with a greater number of tabular points (N) than a more powerful but also more complicated formula which can give the same accuracy with a smaller N. The formulae of Gaussian quadrature utilize the selection of locations of the x_j's as well as the coefficients w_j to give very efficient quadrature formulae if one is interested only in evaluating $\int K(x)\ f(x)\ \mathrm{d}x$ given $f(x)$ and a weight function $K(x)$; but the choice of x_j depends on the weight function, so if Gaussian quadrature was applied to approximate $\int K_i(x)\ f(x)\ \mathrm{d}x$ by the sum $\Sigma_j w_{ij}\ f(x_j)$, the x_j's would be different for each i and a tractable formula would not result. We will content ourselves therefore with some quite elementary quadrature formulae using prescribed, rather than variable, x_j's — for, although elementary, these are capable of any required accuracy in physical problems where the functions $K_i(x)$ and $f(x)$ are usually well-behaved; all that is needed to produce a given accuracy is a sufficiently dense spacing of tabular points, but it is rare indeed for any more than at most 20—30 points to be required. The quadrature formula given in Chapter 2 (section 2.2) is sufficient for most purposes involving the reasonably well-behaved functions which crop up in real problems. In the absence of appreciable high-frequency components quadrature is a fairly simple matter and as we have stated repeatedly it is the absence of high-frequency components in the kernels $K_i(x)$ which gives rise to many of the problems of inversion mathematics. The construction of accurate quadrature formulae is not a source of difficulties. The quadrature formula given in section 2.2. is not much more complicated than the simple linear fit formula from which it is derived. Had a weight function $w(x)$ been included in the integral only the definite integrals would have to be replaced — instead of $x_m - x_{m-1}$ and $\frac{1}{2}(x_m^2 - x_{m-1}^2)$ one would have $\int_{x_{m-1}}^{x_m} w(x)\,dx$ and $\int_{x_{m-1}}^{x_m} xw(x)\,dx$, respectively.

By proceeding a little more circumspectly and using matrix-vector notation the process can be made much tidier. The linear fit formula can be written:

$$(x_m - x_{m-1}) f(x) = [1 \quad x] \begin{bmatrix} x_m & -x_{m-1} \\ -1 & 1 \end{bmatrix} \begin{bmatrix} f_{m-1} \\ f_m \end{bmatrix}$$

in terms of a 2×2 matrix and vectors, or as a N-dimensioned relationship:

$$f(x) = [1 \quad x] \begin{bmatrix} 0 & 0 & 0 \ldots x_m & -x_{m-1} & \ldots 0 \\ 0 & 0 & 0 \ldots -1 & 1 & \ldots 0 \end{bmatrix} \begin{bmatrix} f_1 \\ f_2 \\ \cdot \\ \cdot \\ f_{m-1} \\ f_m \\ \cdot \\ \cdot \\ f_N \end{bmatrix} \cdot (x_m - x_{m-1})^{-1}$$

$(1 \times 1) \quad (1 \times 2) \quad (2 \times N) \quad (N \times 1)$

So:

$$\int_{x_{m-1}}^{x_m} w(x) f(x) \mathrm{d}x = (x_m - x_{m-1})^{-1} \left[\int_{x_{m-1}}^{x_m} w(x) \mathrm{d}x \quad \int_{x_{m-1}}^{x_m} x w(x) \mathrm{d}x \right]$$

$$\begin{bmatrix} 0 & 0 & 0 \ldots x_m & -x_{m-1} & 0 \ldots 0 \\ 0 & 0 & 0 \ldots -1 & 1 & 0 \ldots 0 \end{bmatrix} \begin{bmatrix} f_1 \\ f_2 \\ \cdot \\ \cdot \\ f_N \end{bmatrix}$$

By calculating explicitly the product of the (1×2) row vector and the $(2 \times N)$ matrix and summing over all values of m from 1 to $N-1$, we obtain a $1 \times N$ array which is the array of quadrature coefficients.

This procedure can be applied just as easily to more complicated curve-fitting formulae. The key point is that whether we fit line segments, parabolic segments or higher order curves the coefficients are *linear combinations* of $f_1, f_2, ..., f_N$; if the fit is by least squares or by Fourier methods that still holds true. The integration process is completed by evaluating integrals, multiplying them by the fitting coefficients and summing the result over all intervals; the end result is a linear combination of the coefficients and therefore of the ordinates $f_1, f_2, ..., f_N$.

To illustrate the point consider the quadratic case. We wish to find a quadratic $c_0 + c_1x + c_2x^2$ which passes through $f(x_{m-1}), f(x_m)$ and $f(x_{m+1})$ at $x = x_{m-1}, x_m$ and x_{m+1}, respectively. That is:

$$\begin{bmatrix} f_{m-1} \\ f_m \\ f_{m+1} \end{bmatrix} = \begin{bmatrix} 1 & x_{m-1} & x_{m-1}^2 \\ 1 & x_m & x_m^2 \\ 1 & x_{m+1} & x_{m+1}^2 \end{bmatrix} \begin{bmatrix} c_0 \\ c_1 \\ c_2 \end{bmatrix}$$

So if we denote the inverse of the above 3×3 matrix by $\mathbf{B}$, the fitting formula is:

$$f(x) \cong [1 \quad x \quad x^2] \begin{bmatrix} \\ \mathbf{B} \\ \\ \end{bmatrix} \begin{bmatrix} f_{m-1} \\ f_m \\ f_{m+1} \end{bmatrix}$$

and the integral $\int wf(x)\,dx$ is obtained by inserting the row vector $(\int w\,dx, \int wx\,dx, \int wx^2\,dx)$ in place of $(1, x, x^2)$. The product of the first two terms gives a 1×3 matrix which can be filled with zeros into a $1 \times N$ matrix; summation over all intervals again completes the process.

The computation of a set of quadrature formulae involving the same abscissae but different weight functions amounts to the computation of a matrix $\mathbf{A}$ such that $\Sigma_j\, a_{ij}\, f(x_j) \cong \int K_i(x)\, f(x)\,dx$; this requires the repeated computation (once for each i) of quadrature formulae, $K_i(x)$ being treated as a weight function. The integrals $\int K_i(x)\,dx$, $\int xK_i(x)\,dx$, etc., may be analytically tractable (as, for example, when $K_i(x)$ is $e^{-y_i x}$ or $e^{-iy_i x}$) or they may themselves require integration by numerical (quadrature) methods.

Quadrature formulae of course need not always be based on polynomials or on an exact fit at a number of points equal to the number of coefficients required (e.g. 2 for a linear fit, 3 for a quadratic, and so on). The function $f(x)$ can be fitted by arbitrary functions, say $\psi_1(x), \psi_2(x), ..., \psi_p(x)$ and the coefficients determined for exact fit at p points $x_{m-p+1}, x_{m-p+2}, ..., x_{m-1}, x_m$. Alternately, more than p points can be used and a least-squares fit prescribed at those l points. The least-square fit in turn can be least squares with respect to summation or integration, but the solution to the problem in any of these ramifications is:

$$f(x) \cong b_1 \psi_1(x) + b_2 \psi_2(x) + ... + b_p \psi_p(x)$$

the array of b's being given by:

$$\begin{bmatrix} b_1 \\ b_2 \\ \cdot \\ \cdot \\ b_p \end{bmatrix} = \begin{bmatrix} \psi_1 \cdot \psi_1 & \psi_1 \cdot \psi_2 \; ... \\ \psi_2 \cdot \psi_1 & \psi_2 \cdot \psi_2 \; ... \\ \cdot & \cdot \quad\quad \cdot \\ \cdot & \cdot \quad\quad \cdot \\ \cdot & \cdot\cdot \quad\quad \cdot \end{bmatrix}^{-1} \begin{bmatrix} \psi_1 \cdot f \\ \psi_2 \cdot f \\ \cdot \\ \cdot\cdot \\ \psi_p \cdot f \end{bmatrix}$$

the dot product being computed according to whatever square norm is to be minimized in the least-squares operation.

Even in the realms of polynomials there are many choices — a quadratic can be fitted by least squares at four points and used for integration between the middle pair, a cubic can be made to fit exactly at x_m and the other three degrees of freedom assigned by least squares on four neighboring points, and so on. The usefulness of such procedures must be judged according to the particular problem, but in the writer's experience complicated quadrature formulae rarely produce a marked improvement in inversion work. When some thought is given to the location of the quadrature points (the tabular values x_1, x_2, etc.) and to the scaling of a problem and the choice of variables it is seldom necessary to go beyond the simple trapezoidal and Simpson's rule formulae. The complicated formulae, however, still lead to expressions which are linear in $f_1, f_2, ..., f_N$ and give:

$$g = \int k(x) f(x) \mathrm{d}x \cong \mathbf{A}f$$

just as the simpler formulae do.

We will not go further into the details of the many quadrature formulae which exist and the many more which can be derived. The interested reader will find quite a number of formulae and a stimulating discussion of quadrature errors in Lanczos' (1956) text "Applied Analysis".

5.5. THE FOURIER TRANSFORM AND FOURIER SERIES

In many physical problems it is desirable to find a Fourier series:

$$a_0 + a_1 \cos x + a_2 \cos 2x + ... + b_1 \sin x + b_2 \sin 2x + ...$$

to represent a given function, or to seek a solution for a problem in terms of a Fourier series. We will not go into the question of the validity in an analytic sense of such a representation — the interested reader will find many discussions of this question in texts on analysis and on Fourier series and Fourier integrals — such a representation will be assumed to be legitimate. There are very few functions and no well-behaved functions which cannot be approximated arbitrarily closely by a Fourier series.

Since $\cos x$, $\sin x$, and all higher-order sinusoids repeat themselves exactly every 2π interval in x, the interval in x if not 0 to 2π or $-\pi$ to π is rescaled to one or the other of these intervals. The Fourier series can then be obtained using the orthogonality of the trigonometric functions on those intervals. In fact, if:

$$a_0 = \frac{1}{2\pi}\int_0^{2\pi} f(x)\,dx, \quad a_k = \frac{1}{\pi}\int_0^{2\pi} \cos kx\, f(x)\,dx, \quad b_k = \frac{1}{\pi}\int_0^{2\pi} \sin kx\, f(x)\,dx$$

then:

$$f(x) = a_0 + \sum_{k=1}^{\infty} (a_k \cos kx + b_k \sin kx)$$

and, as we have seen, if we truncate the sum after m terms, the approximation:

$$f(x) \cong a_0 + \sum_{k=1}^{m} (a_k \cos kx + b_k \sin kx) = f_m(x)$$

is optimal with respect to the residual, $\int_0^{2\pi} [f(x) - f_m(x)]^2\, dx$, which is minimized by the a_k, b_k given above.

It is not necessary that $f(x)$ be continuous to obtain a Fourier series, but it must be defined on an interval of width 2π. To rescale linearly another interval (a, b) to $(-\pi, \pi)$ is a unique process and the Fourier series obtained is unique; if the scaling is to $(0, 2\pi)$ we obtain another unique Fourier series which is trivially different to the former, differing only to the extent that $\cos(\theta + \pi) = -\cos\theta$, $\sin(\theta + \pi) = -\sin\theta$. But it is also possible to scale (a, b) to, say, $(0, \pi)$ and prescribe that $f(x)$ behave in some quite arbitrary way in the remaining part of a 2π interval. For example one could prescribe any one of the following:

(a) $f(x) \equiv f(\pi)$ from π to 2π

(b) $f(x) \equiv f(0)$ from 0 to $-\pi$

(c) $f(x) = 0$ from π to 2π

(d) $f(x + \pi) = f(x)$ from $x = 0$ to $x = \pi$

(e) $f(x) = f(-x)$ from $x = 0$ to $x = -\pi$

and so on. These will give different Fourier series, differing in form and in relative accuracy after m terms, but summed to infinity they will *all* converge to $f(x)$ in the interval $(0, \pi)$ and outside this interval will exhibit the behavior prescribed in the recipe whereby the interval was extended. Hence for a finite interval (a, b), we can rescale to an interval of width 2π or less, arbitrarily postulate a behavior in the remaining part of the 2π and thereby obtain widely different Fourier series.

Harmonic analysis

If $f(x)$ is tabulated at $2N + 1$ evenly spaced points, $x_{-N}, x_{-N+1}, \ldots, x_N$, we can use the summation orthogonality of the trigonometric functions to produce coefficients $a_0, a_1, \ldots, a_N$ and $b_1, b_2, \ldots, b_N$, such that the "trigonometric polynomial":

$$a_0 + \sum_{1}^{N} (a_k \cos kx + b_k \sin kx)$$

takes precisely the values $f(x_1), f(x_2)$, etc., at the points $x_{-N}, x_{-N+1}, \ldots, x_N$. If $x_j = j \cdot 2\pi/(2N + 1)$, the results obtained earlier in this chapter give immediately a_k and b_k:

$$a_k = \frac{2}{2N+1} \sum_{j=-N}^{N} \cos kx_j\, f(x_j) \qquad (k \neq 0)$$

$$b_k = \frac{2}{2N+1} \sum_{j} \sin kx_j\, f(x_j)$$

$$a_0 = \frac{1}{2N+1} \sum_{j} f(x_j)$$

The a_k and b_k are not the same as those of the Fourier series for $f(x)$ on the interval $(-\pi, \pi)$. The expression for a_k for the Fourier series is:

$$a_k = \frac{1}{\pi} \int_{-\pi}^{\pi} \cos kx\, f(x) dx$$

This can certainly be approximated for well-behaved $f(x)$ by a sum Σ_j $\cos kx_j\ f(x_j)\ \Delta x$ and since $\Delta x = 2\pi/(2N + 1)$; the expression in the trigonometric interpolation differs from the Fourier series expression only to the extent that a histogram-type summation formula (i.e. the very crudest quadrature) replaces an integral. Evidently in the limit $2N + 1 \to \infty$ the two sets of coefficients become the same (provided only that the integrals, as defined by the elementary summation limit definition, exist).

The Fourier integral

If $f(x)$ is defined on the interval $(-a, a)$, this can readily be scaled to $(-\pi, \pi)$, using the transformation $\theta = \pi x/a$. An N-order trigonometric polynomial can be fitted to any $f(x)$ and it will be exact at the points $\theta_j = [(-N + j)/(2N + 1)]2\pi$. The resulting interpolation is:

$$\sum_{k=-N}^{N} c_k e^{ik\theta}$$

if for compactness and symmetry we write, utilizing complex notation:

$$a_k \cos k\theta + b_k \sin k\theta = \tfrac{1}{2}(a_k - ib_k)\,e^{ik\theta} + \tfrac{1}{2}(a_k + ib_k)\,e^{-ik\theta}$$

and allow k (hitherto positive) to take negative values. For positive k, $c_k = \frac{1}{2}(a_k - ib_k)$ while for negative k it takes the value $\frac{1}{2}(a_k + ib_k)$. The formulae for a_k and b_k already derived for trigonometric interpolation:

$$a_k = \frac{2}{2N+1}\sum_{j=-N}^{N} \cos k\theta_j\, f(x_j)$$

$$b_k = \frac{2}{2N+1}\sum_{j=-N}^{N} \sin k\theta_j\, f(x_j)$$

are evidently equivalently to the complex formula:

$$c_k = \frac{2}{2N+1}\sum_{j=-N}^{N} (\cos k\theta_j - i \sin k\theta_j)\, f(x_j)$$

or:

$$c_k = \frac{2}{2N+1}\sum_{j=-N}^{N} \exp(-i\pi k x_j/a)\, f(x_j)$$

while the interpolation for $f(x)$, an interpolation which we have seen to be exact at the $2N + 1$ tabular points, is:

$$f(x) \cong \sum_{k=-N}^{N} c_k\, e^{i\pi kx/a}$$

The similarity and symmetry of the last two formulae is obvious. In present-day terminology the process of multiplication of equally spaced values of $f(x)$ by $\exp(ikx_j\pi/a)$ is often called a Fourier transform; in this terminology the results just given establish rigorously the remarkable property of such transforms, that two applications of the transform operations restores the original function (more strictly vector), apart from a possible scalar factor.

However, the original classical definition of a Fourier transform is not a sum, but an integral. The Fourier transform of $f(x)$ is $\int_{-\infty}^{\infty} e^{i\omega x} f(x)\,dx$, a function $s(\omega)$ of the continuous variable ω, and the Fourier integral theorem states essentially that a second such operation restores the original function, apart from a scalar factor which may be removed if the Fourier transform is defined as $(1/\sqrt{2\pi}) \int_{-\infty}^{\infty} e^{i\omega x} f(x)\,dx$. The integral theorem may be written:

If

$$s(\omega) = \frac{1}{\sqrt{2\pi}} \int_{-\infty}^{\infty} e^{i\omega\xi} f(\xi)\,d\xi$$

then

$$f(x) = \frac{1}{\sqrt{2\pi}} \int_{-\infty}^{\infty} e^{-i\omega x} s(\omega)\,d\omega$$

a remarkable and unique relationship. It is not appropriate to attempt to establish the integral theorem in a rigorous and most general way here. Suffice it to point out that the expression derived for c_k above tends to an integral of the form $a^{-1} \int_{-a}^{a} e^{-i\omega x} f(x)\,dx$ as N tends to infinity while the interpolation approximation for $f(x)$ tends to an integral of the form πa^{-1} $\int_{-(N+1/2)\pi/a}^{(N+1/2)\pi/a} e^{i\omega x} c(\omega)\,d\omega$, when ω is identified with $\pi k/a$ and $c(\omega)$ is c_k, regarded as a function of the continuous variable ω. These arguments do not prove the integral theorem, but serve to show its reasonableness and prove the validity of the theorem to the extent that $f(x)$ can be replaced for large enough N by the interpolating trigonometric polynomial.

The Fourier integral theorem can be applied for a very wide range of types of functions, not necessarily continuous or bounded. The various formulae for Fourier series and trigonometric polynomials are contained within it as special cases and it can be used to obtain integral representations for non-analytic "functions" such as the delta function or Dirichlet's discontinuous function which vanishes outside the interval ± 1, and is unity within it.

We can therefore represent any function by either a finite or infinite sum of discrete trigonometric components or in the most general case an integral over all frequencies ω. If the function is smooth, the spectrum (i.e. $s(\omega)$ or the discrete coefficients c_m, a_m, b_m) will fall off quickly with increasing frequency (order). Narrow functions or functions with discontinuities give spectra which fall off more slowly. Consider, for example, a discontinuous function taking a constant value from $-h/2$ to $+h/2$; if we make the constant value $1/h$ the area under this function is unity. Its Fourier transform is:

$$h^{-1} \int_{-h/2}^{h/2} e^{i\omega x}\,dx = \frac{\sin \omega h/2}{\omega h/2}$$

This is an oscillating function of ω, diminishing in amplitude as ω increases because of the ω^{-1} term. The first zero is located at $\omega = 2\pi/h$, the second at $4\pi/h$, and so on, and as h increases these, and all other points on the curve, move in towards the origin as h^{-1}. There is therefore an inverse relationship between the width of a function and the width of its spectrum (i.e. its Fourier transform, or its Fourier coefficients plotted against their order).

The aspect of smoothness of $f(x)$ and its effect on the spectrum can be viewed in another way. Since we are mainly concerned with finite intervals in x, we will employ Fourier series, but this is not material to the result. From [5.7] we have:

$$a_m = \int_{-\pi}^{\pi} \cos mx\, f(x)\mathrm{d}x$$

Now if $f(x)$ is differentiable, we can integrate by parts:

$$a_m = -\frac{1}{m}\int_{-\pi}^{\pi} \sin mx\, f'(x)\mathrm{d}x$$

(f' here signifies $\mathrm{d}f/\mathrm{d}x$). If $f'(x)$ is differentiable, integration by parts can be repeated:

$$a_m = \frac{1}{m^2} \cos mx\, f'(x)|_{-\pi}^{\pi} - \frac{1}{m^2}\int_{-\pi}^{\pi} \cos mx\, f''(x)\mathrm{d}x$$

It is very common for $f'(x)$ to vanish or take fixed values at the end points, so that the spectrum very often falls off effectively as m^{-2} or even faster. One is therefore very dependent on discontinuities or kinks in $f(x)$ to limit the rate at which the spectrum of $f(x)$ falls off with frequency.

The Faltung (convolution) theorem

In an earlier chapter the convolution ("Faltung") product of two functions $f(x)$ and $g(x)$ was mentioned and its physical relevance to smearing or blurring processes was referred to. The convolution of $f(x)$ and $g(x)$ was defined as:

$$f^*g = \int_{-\infty}^{\infty} f(\xi)g(x-\xi)\mathrm{d}\xi$$

it is a function of x, and clearly additive, distributive and associative as far as f and g are concerned. The Fourier transform of the convolution of f and g is related to the Fourier transforms of f and g separately in a simple and important way. It is by definition:

$$\int_{-\infty}^{\infty} e^{i\omega x} \int_{-\infty}^{\infty} f(\xi)g(x-\xi)\mathrm{d}\xi\,\mathrm{d}x$$

Provided interchange of the order of integration is permissible, this double integral can be written:

$$\int_{-\infty}^{\infty} f(\xi)e^{i\omega\xi} \int_{-\infty}^{\infty} e^{i\omega(x-\xi)}g(x-\xi)\mathrm{d}x\,\mathrm{d}\xi$$

or:

$$\int_{-\infty}^{\infty} f(\xi)e^{i\omega\xi}\mathrm{d}\xi \int_{-\infty}^{\infty} g(\xi')e^{i\omega\xi'}\mathrm{d}\xi'$$

the product of the Fourier transform of $f(x)$ and the Fourier transform of $g(x)$. This is the Faltung theorem.

Since the Fourier transform of the Fourier transform of a function gives the original function (apart from a possible numerical factor), the theorem can also be stated conversely as follows: the Fourier transform of the algebraic product $f(x)\,g(x)$ is the convolution of the separate Fourier transforms $\int_{-\infty}^{\infty} e^{i\omega x} f(x)\,\mathrm{d}x$ and $\int_{-\infty}^{\infty} e^{i\omega x} g(x)\,\mathrm{d}x$; this latter convolution is of course obtained by an integration over the frequency ω. The convolution theorem has direct relevance to inversion problems and to the integral equation:

$$g(y) = \int K(y,x)f(x)\mathrm{d}x$$

which in many situations can be written as a convolution of K and f. The spectrum of g will be band-limited (i.e. will be zero or negligible above some maximum ω_{max} if *either* K or f is.

5.6. SPECTRAL FORM OF THE FUNDAMENTAL INTEGRAL EQUATION OF INVERSION

The Fourier transform of a function of time gives the frequency spectrum of the function, and in a natural extension of the terminology it is convenient to refer to the function $\int_{-\infty}^{\infty} e^{i\omega x} f(x)\,\mathrm{d}x$ as the spectrum of $f(x)$ or the spectral function associated with $f(x)$. The integral theorem implies that either function is fully equivalent to the other. Since $f(x)$ can be written $\int_{-\infty}^{\infty} s(\omega)\,e^{-i\omega x}\,\mathrm{d}\omega$, the fundamental Fredholm integral equation of the first kind can be written:

$$g(y) = \int_{a}^{b}\int_{-\infty}^{\infty} s(\omega)K(y,x)e^{-i\omega x}\mathrm{d}\omega\,\mathrm{d}x$$

or, reversing the order of integration and defining $\Phi(y, \omega)$ as $\int_a^b K(y, x)\, e^{-i\omega x}\, dx$:

$$g(y) = \int_{-\infty}^{\infty} \Phi(y, \omega) s(\omega) d\omega$$

again a Fredholm integral equation, with the (unknown) spectral function $s(\omega)$ appearing in place of the unknown function $f(x)$. Since $K(y, x)$ is by its physical nature smooth, the infinite interval can in practice often be replaced by a finite one. Moreover, the smoothness of physical kernels has a particular significance as far as $\Phi(y, \omega)$ is concerned, for as was briefly discussed earlier smooth functions give a spectrum which falls off quickly — as $|\omega|^{-2}$ or even $|\omega|^{-3}$ — with increasing $|\omega|$. The spectrum may or may not oscillate initially but eventually it falls off quickly with increasing frequency. The situation for smooth kernels $K(y, x)$ is as shown in Fig. 5.2 — the functions $s(\omega)$ are "blind" to higher frequencies and high frequency components in the solution are inaccessible. Even if we have a great number of kernels the higher-frequency components cannot be retrieved, just as a large number of measurements through different red filters (however numerous) do not enable one to establish the blue or ultraviolet part of an optical spectrum.

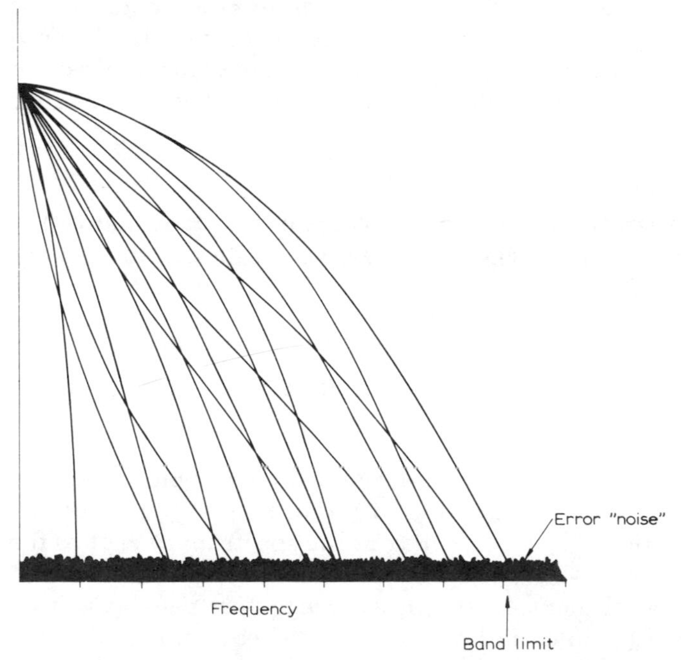

Fig. 5.2. Schematic diagram illustrating the limited number of pieces of information which can be obtained in the presence of errors from a "band-limited" set of kernels (even though the set may be very large in number).

BIBLIOGRAPHY

Arfken, G., 1970. *Mathematical Methods for Physicists.* Academic Press, New York, N.Y., 815 pp.

Båth, M., 1968. *Mathematical Aspects of Seismology.* Elsevier, Amsterdam, 416 pp.

Bocher, M., 1960. *An Introduction to the Study of Integral Equations.* Hafner, New York. N.Y., 71 pp.

Carslaw, H.S., 1930. *Theory of Fourier Series and Integrals.* Oxford University Press, Oxford, 368 pp.

Courant, R. and Hilbert, D., 1953. *Methods of Mathematical Physics.* Interscience, New York, N.Y., 561 pp.

Lanczos, C., 1956. *Applied Analysis.* Prentice-Hall, Englewood Cliffs, N.J., 539 pp.

National Research Council, 1940. Bibliography of orthogonal polynomials. *Bull Natl. Res. Counc. (U.S.),* 103, 204 pp.

Papoulis, A., 1962. *The Fourier Integral and Its Applications.* McGraw-Hill, New York., N.Y., 318 pp.

Szego, G., 1939. *Orthogonal Polynomials.* American Mathematical Society, New York, N.Y., 401 pp.

Titchmarch, E.C., 1948. *Introduction to the Theory of Fourier Integrals.* Oxford University Press, Oxford, 394 pp.

Whittaker, E.T. and Watson, G.N., 1962. *A Course in Modern Analysis.* Cambridge University Press, Cambridge, 608 pp.

CHAPTER 6

Linear inversion methods

The fundamental integral equation of indirect measurement, $g(y) = \int K(y, x)\, f(x)\, dx$, is a linear equation. The formal inverse is also linear and many (but not all) inversion algorithms are linear, in the sense that if a certain $g_1(y)$ gives a solution $f_1(x)$ and $g_2(y)$ a solution $f_2(x)$ then $g_1(y) + \beta g_2(y)$ gives the solution $f_1(x) + \beta f_2(x)$. Many of these linear methods are very similar to the inversion of systems of linear equations and thus differ from the procedures applied to simple puzzles such as those mentioned in Chapter 2, only because of the necessity to apply in some way constraints to the solution to prevent the appearance of the instabilities which have been discussed from several aspects and at some length in the preceding chapters.

6.1. QUADRATURE INVERSION

Examples of quadrature formulae have been given already. These formulae are simply methods for replacing integrals $\int_a^b w(x)\, u(x)\, dx$ by sums $\Sigma_j\, c_j\, u(x_j)$ where the quadrature coefficients c_j depend on the interval (a, b), on the number and location of the x_j's and the weight function $w(x)$. The kernels $K(y_i, x)$ can be regarded as a weighting function which varies with y_i and so gives a set of coefficients for each i. Thus:

$$g(y_1) = c_1(y_1)\, f(x_1) + c_2(y_1)\, f(x_2) + \ldots + c_N(y_1)\, f(x_N)$$

$$g(y_2) = c_1(y_2)\, f(x_1) + c_2(y_2)\, f(x_2) + \ldots + c_N(y_2)\, f(x_N)$$

$$\cdot$$

$$\cdot$$

$$g(y_M) = c_2(y_M)\, f(x_1) + c_2(y_M)\, f(x_2) + \ldots + c_N(y_M)\, f(x_N)$$

or, if we define a matrix $\mathbf{A} = \| c_j(y_i) \|$ and define vectors $\boldsymbol{g}$ and $\boldsymbol{f}$ in the obvious way:

$$\boldsymbol{g} = \begin{bmatrix} g(y_1) \\ g(y_2) \\ \cdot \\ \cdot \\ g(y_M) \end{bmatrix} \qquad \boldsymbol{f} = \begin{bmatrix} f(x_1) \\ f(x_2) \\ \cdot \\ \cdot \\ f(x_N) \end{bmatrix}$$

then:

$$g \cong \mathbf{A} f$$

or:

$$\mathbf{A} f = g + \epsilon \qquad [6.1]$$

where ϵ is an error due to the quadrature.

At this point a slight digression is in order to emphasize the important fact that the error in the quadrature operation is not large when the functions involved are reasonably smooth and a sufficient number of tabular values x_1, x_2, ..., x_N are used and they are located judiciously. In Table 6.1 are listed for a number of values of y the integral $\int_0^1 x\, e^{-yx} f(x)\, dx$ for the function $f(x) = 1 + 4(x - \frac{1}{2})^2$; the integral was evaluated analytically to obtain the

TABLE 6.1

Values of $\int_0^1 x\, e^{-yx} f(x)\, dx$ computed by quadrature and exactly

y	Exact value	Quadrature	Error
0.10	0.6218	0.6226	0.0008
0.20	0.5803	0.5810	0.0007
0.30	0.5419	0.5426	0.0007
0.40	0.5063	0.5070	0.0007
0.50	0.4734	0.4740	0.0006
0.60	0.4430	0.4435	0.0005
0.70	0.4147	0.4153	0.0006
0.80	0.3886	0.3891	0.0005
0.90	0.3643	0.3648	0.0005
1.0	0.3418	0.3422	0.0004
1.2	0.3015	0.3019	0.0004
1.4	0.2667	0.2670	0.0003
1.6	0.2366	0.2369	0.0003
1.8	0.2106	0.2109	0.0003
2.0	0.1880	0.1882	0.0002
2.5	0.1434	0.1436	0.0002
3.0	0.1116	0.1117	0.0001
3.5	0.0885	0.0886	0.0001
4	0.0714	0.0715	0.0001
5	0.0490	0.0490	0.0000
6	0.0356	0.0356	0.0000
7	0.0271	0.0271	0.0000
8	0.0214	0.0214	0.0000
9	0.0173	0.0173	0.0000
10	0.0144	0.0144	0.0000

values in the column headed "exact", while the column labeled "quadrature" contains values computed by the trapezoidal formula equation [2.3] (a formula which is exact only when $f(x)$ is linear between successive tabular values). Twenty tabular values were used. The data calculated by the quadrature formula is therefore closer to the exact values than most physical measurements are capable of achieving. The maximum relative error in the above data is 0.14% and the overall root-mean-square (r.m.s.) error is 0.11%. This sort of quadrature accuracy is typical of what can be achieved in most practical indirect sensing experiments and it serves to illustrate the point that while quadrature does produce an error, the quadrature error is not usually the dominant contributor to the overall uncertainty.

Returning now to the inversion itself — we can use the data such as that calculated and displayed in Table 6.1 as a hypothetical input for the problem of estimating $f(x)$ (which we know to be $1 + 4(x - \frac{1}{2})^2$) from the tabulated set of values of $g(y_i) = \int K_i(x)\, f(x)\, dx$. The quadrature which we carried out (trapezoidal interpolation to $f(x)$, the kernel $x\, e^{-yx}$ treated as a weighting function and the resulting integrals evaluated analytically) as we have seen earlier can be represented by a matrix of quadrature coefficients which premultiplies the vector $\boldsymbol{f}$ to arrive at the vector $\boldsymbol{g}$. We may attempt the inversion using the exact values tabled above or the quadrature values.
From:

$$\mathrm{A}\boldsymbol{f} = \boldsymbol{g} + \epsilon$$

it is obvious step to a solution:

$$\boldsymbol{f}' = \mathrm{A}^{-1}\boldsymbol{g} \qquad [6.2]$$

(using the prime to indicate a solution — not a derivative); we have calculated the results of this combination using both "exact" and "quadrature" values for $\boldsymbol{g}$. The results are disappointing as Table 6.2 shows.

Both results are very bad when compared with the original $f(x)$, yet the values obtained for $\boldsymbol{g}$ when these unacceptable solutions are inserted in place of $f(x)$ are *very close* indeed to the original $\boldsymbol{g}$ — in fact they are closer than most measurements, even those of the highest quality, would be. Table 6.3 shows how close $\int K_i(x)\, f'(x)\, dx$ is to the original g_i's. It is important to note that the inversion of the quadrature-computed values using the matrix A computed by the same quadrature still is not close to the original $f(x_i)$ — this of course arises from small roundoff errors in the computation of $\boldsymbol{g}$ which can be amplified by the inversion; by definition $\mathrm{A}^{-1}\, \boldsymbol{g}$ must be exactly $\boldsymbol{f}$ if $\boldsymbol{g}$ is exactly $\mathrm{A}\boldsymbol{f}$ and if the computations were mathematically exact the inversion must return $\boldsymbol{f}$.

Instability

One's first reaction to the failure of direct inversion is usually to look for more accuracy — to seek better quadrature formulae, to use double-precision

TABLE 6.2

Result of direct inversion $f' = \mathbf{A}^{-1}\mathbf{g}$ on data of Table 6.1 compared with the $f(x)$ from which the data was computed

i	x_i	$f(x_i)$	f'_i from "exact" $\mathbf{g}$	f'_i from quadrature $\mathbf{g}$
1	0.05	1.81	7979	2280
2	0.1	1.64	−26,222	−7287
3	0.15	1.49	48,215	12,860
4	0.2	1.36	−41,271	−9589
5	0.25	1.25	−11,197	6380
6	0.30	1.16	41,763	15,276
7	0.35	1.09	5160	−1504
8	0.4	1.04	35,417	−10,588
9	0.45	1.01	−20,377	2487
10	0.5	1.00	39,681	1265
11	0.55	1.01	24,757	6342
12	0.6	1.04	−39,393	−4331
13	0.65	1.09	−11,035	−2797
14	0.7	1.16	41,353	10,929
15	0.75	1.25	−61,988	−25,922
16	0.8	1.36	67,292	23,292
17	0.85	1.49	−21,011	4742.1
18	0.9	1.64	−23,273	−20,714
19	0.95	1.81	19,687	11,774
20	1.00	2.00	−1158	−318.2

arithmetic, and so on. There is, however, little point to that, for what we have found is not merely imprecision — it is *instability*. Small changes in $\mathbf{g}$, as from the "exact" to the "quadrature" data, produce large changes in the

TABLE 6.3

Original g_i compared to that given by the solution f'

$\int K_i(x)\, f(x)\, dx$	g_i	$\int K_i(x)\, f'(x)\, dx$	g_i
0.00412	0.00415	0.2092	0.2087
0.01427	0.01440	0.2284	0.2278
0.02936	0.02995	0.2464	0.2457
0.04887	0.04903	0.2632	0.2625
0.07142	0.07152	0.2789	0.2781
0.09550	0.09551	0.2936	0.2927
0.1199	0.1198	0.3072	0.3063
0.1438	0.1436	0.3201	0.3200
0.1668	0.1665	0.3321	0.3318
0.1886	0.1882	0.3433	0.3422

solution f' and the situation is totally untenable when measurements are going to be used to get values for the g_i, for then it is totally unrealistic to ask for "exactness". When we measure g_i we merely establish a likelihood of the actual g_i lying in a certain, necessarily finite, interval.

It is instructive to look at the elements of $\mathbf{A}^{-1}$. There are several hundred of them, so we will look at only a few — the first couple of rows of $\mathbf{A}^{-1}$. These are set out in Table 6.4. There are obviously some very large numbers among the elements of $\mathbf{A}^{-1}$. The presence of such very large elements must inevitably produce instability: a small change Δg_{10} in g_{10}, for example, leads to a change $\alpha_{1,10}\Delta g_{10}$ in f_1 and a change $\alpha_{2,10}\Delta g_{10}$ in f_2, or numerically $6.9 \times 10^{12}\Delta g_1$ and $-2.3 \times 10^{13}\Delta g_1$. *The error is enormously magnified.*

Since it is impossible to invert a singular matrix, it seems likely that near-singular matrices may be difficult to invert and the computed inverses inaccurate to some degree. It is indeed true that the near-singular matrices of indirect sensing are more difficult to invert, but it is important to understand clearly that this is a symptom of near-singularity and not a cause of the instability. If $\mathbf{A}$ has a small eigenvalue, its inverse will have a large eigenvalue and inevitably will thereby contain some large elements. It will also be more difficult to compute $\mathbf{A}^{-1}$ explicitly to some prescribed accuracy, but it is the large eigenvalues and elements of $\mathbf{A}^{-1}$ that constitute the fundamental dif-

TABLE 6.4

Elements of first two rows of $\mathbf{A}^{-1} = \|\alpha_{ij}\|$

j	Row 1 (α_{1j})	Row 2 (α_{2j})
1	2.3×10^5	-6.4×10^5
2	-2.3×10^7	7.1×10^7
3	1.0×10^9	-3.1×10^9
4	-1.9×10^{10}	6.1×10^{10}
5	1.8×10^{11}	-5.9×10^{11}
6	-9.6×10^{11}	3.1×10^{12}
7	2.6×10^{12}	-8.4×10^{12}
8	-3.2×10^{12}	1.0×10^{13}
9	-1.2×10^{12}	4.3×10^{12}
10	6.9×10^{12}	-2.3×10^{13}
11	-4.1×10^{12}	1.4×10^{13}
12	-3.3×10^{12}	1.1×10^{13}
13	3.2×10^{12}	-1.0×10^{13}
14	3.4×10^{11}	-1.1×10^{12}
15	1.1×10^{11}	-4.1×10^{11}
16	2.9×10^{11}	-9.7×10^{11}
17	-1.3×10^{12}	4.3×10^{12}
18	2.0×10^{10}	-1.4×10^{10}
19	6.3×10^{11}	-2.1×10^{12}
20	-1.7×10^{11}	5.7×10^{11}

ficulty and these are results of nothing more than the existence of small eigenvalues of **A**, which in turn result from a high degree of interdependence among some of the kernels $K_i(x)$. If there is only a very small difference between some particular kernel and a linear combination of the others, then the same situation will prevail among the rows of **A** and since a determinant is unaltered by substracting from any row any linear combination of other rows, the determinant det(**A**) can be transformed into one with very small elements in all positions of one row, so that det(**A**) will be very small; the determinant, as we have seen, equals the product of the eigenvalues of **A**, so small eigenvalues are another manifestation of the same condition-interdependence of kernels.

6.2. LEAST SQUARES SOLUTION

One's next reaction to instability is often to seek more data, to argue that if 20 measurements of g_i cannot satisfactorily give 20 tabular values of $f(x_i)$, then perhaps 25, 30, 40 or more can.

The problem of solving such an overdetermined system of equations, wherein there are more equations (M) than unknowns (N), can be readily handled by obtaining a "least squares solution". That is, we find a solution f'' which of all possible vectors of dimension N minimizes the norm of the residual $(\mathbf{A}f'' - g)$ — in other words $\mathbf{A}f''$ lies closest to g in the vector space of M-dimensioned vectors. The square norm, which gauges the magnitude of $(\mathbf{A}f - g)$ can be written:

$$q = (\mathbf{A}f - g)^*(\mathbf{A}f - g)$$
$$= f^*\mathbf{A}^*\mathbf{A}f - g^*\mathbf{A}f - f^*\mathbf{A}^*g + g^*g$$

(we have dropped temporarily the double prime superscripts of f''); the quadratic expression q can be differentiated with respect to f_k: an extremum is attained when $\partial q/\partial f_k = 0$ $(k = 1, 2, ..., N)$, i.e. when:

$$e_k^*\mathbf{A}^*\mathbf{A}f + f^*\mathbf{A}^*\mathbf{A}e_k - e_k^*\mathbf{A}^*g - g^*\mathbf{A}e_k = 0 \qquad (k = 1, 2, ... N)$$

Note that **A** is now an $M \times N$ matrix and $\mathbf{A}f - g$ is an M-dimensioned vector, but there are only N unknowns $f_1, f_2, ..., f_N$, and so N equations, in the above set. e_k is a N-dimensioned vector which has zero in all but the kth element, where a one appears. It is readily seen that the above system of equations is equivalent to the vector equation:

$$\mathbf{A}^*\mathbf{A}f - \mathbf{A}^*g = 0$$

or:

$$\underset{N \times 1}{f''} = \underset{(N \times M \times M \times N)^{-1}}{(\mathbf{A}^*\mathbf{A})^{-1}} \quad \underset{(N \times M)\ (M \times 1)}{\mathbf{A}^*g} \quad \text{dimensions} \qquad [6.3]$$

In [6.3] we have restored the double prime superscript to emphasize the least squares nature of f''. It is useful to note that the symmetric matrix $\mathbf{A}^*\mathbf{A}$ is obtained by forming the scalar products of all pairs of columns of A, the ij and ji elements of $\mathbf{A}^*\mathbf{A}$ being given by the product of the ith and jth columns. Note also that if the number of equations M equals the number of unknown N, A and $\mathbf{A}^*$ become square and f'' is then $\mathbf{A}^{-1}(\mathbf{A}^*)^{-1}\ \mathbf{A}^*g$, i.e. the direct inverse $\mathbf{A}^{-1}g$.

The solution given by the least squares method is no better than that given by the direct inverse — the elements of $(\mathbf{A}^*\mathbf{A})^{-1}$ *tend to be even larger than those of* A *— for we have* $\det(\mathbf{A}^*\mathbf{A}) = (\det \mathbf{A})^2$ *— and this does nothing to improve the situation.*

The literature of inversion contains quite a number of references and discussions of the relative merits of the inversion formulae ([6.2] and [6.3]):

$$f' = \mathbf{A}^{-1}g \tag{6.2}$$

and

$$f'' = (\mathbf{A}^*\mathbf{A})^{-1}\mathbf{A}^*g \tag{6.3}$$

which with exact mathematics are equivalent statements, but do not give the same results in the presence of errors. The eigenvalues of $\mathbf{A}^*\mathbf{A}$ are not explicitly related in any simple way to those of A, but are generally smaller and $(\mathbf{A}^*\mathbf{A})^{-1}$ cannot be computed as accurately as $\mathbf{A}^{-1}$, from this point of view the direct formula [6.2] is preferable. However, if measurement errors in g predominate, it matters little which formula is used. Suppose that when $(\mathbf{A}^*\mathbf{A})^{-1}$ is calculated we in fact have $(\mathbf{A}^*\mathbf{A})^{-1}$ plus an error term which will be written as $\delta\mathbf{Y}$ where Y is a matrix which has been scaled to make its dominant eigenvalue unity and δ is a small scalar quantity. Suppose also that in place of $\mathbf{A}^{-1}$ we have $\mathbf{A}^{-1} + e\mathbf{X}$ and that the error in g is ϵ. Then by formula [6.2] we compute:

$$f' = (\mathbf{A}^{-1} + e\mathbf{X})\,(g + \epsilon)$$

while [6.3] gives:

$$f'' = [(\mathbf{A}^*\mathbf{A})^{-1} + \delta\mathbf{Y}]\,\mathbf{A}^*(g + \epsilon)$$

If we disregard products of the small quantities e, δ and $|\epsilon|$, the errors in both procedures become:

$$\Delta f' = e\mathbf{X}g + \mathbf{A}^{-1}\epsilon$$

$$\Delta f'' = \delta\mathbf{Y}\mathbf{A}^*g + \mathbf{A}^{-1}\,\epsilon$$

Thus if $\mathbf{A}^{-1}\epsilon$ is the dominant error — and it increases with decreasing λ_{min}

and det(**A**) — there is nothing to choose between the two formulae. If the terms in δ and e predominate the direct inversion formula is preferable — but only as the best of a bad lot — since **A** possesses larger eigenvalues than $\mathbf{A}^*\mathbf{A}$ (if **A** happens to be symmetric the eigenvalues of $\mathbf{A}^*\mathbf{A}$ will be the squares of the eigenvalues of **A**). However, the accuracy of an inverse can always be upgraded by iterative procedure and also there are for symmetric matrices such as $\mathbf{A}^*\mathbf{A}$ inversion procedures, notably decomposition into triangular matrices, which essentially "take the square root" of a matrix before inversion and for inversion are not degraded as seriously as the squares of the eigenvalues might suggest.

6.3. CONSTRAINED LINEAR INVERSION

Neither direct nor least squares methods work well in ill-conditioned systems — as we have just seen. Since in either case the root cause is the existence of very small eigenvalues in **A** or $\mathbf{A}^*\mathbf{A}$, the only procedures which can be expected to improve the situation are those which in some way increase the magnitude of the eigenvalues. Since **A** is computed from the kernels associated with the various measurements we can change the eigenvalues only if we change the question which we are asking of the measured data and since error in **g** has been seen to be the other problem element, it is appropriate that the rephrased question should involve the error.

If we make M measurements at $y_1, y_2, ..., y_m$, $g(y)$ or **g** is thereby defined only at those points and to within the measurement error. It is not defined in the sense of Fig. 6.1a. One could in fact define a set of error bars (Fig. 6.1b) or zones (Fig. 6.1c) and say that $g(y)$ is arbitrary except that it passes through each of the bars or zones; the relationship:

$$g(y) = \int_a^b K(y, x)\, f(x)\, dx$$

implies that there exists in the f, x domain a set (probably infinite) of $f(x)$ which are associated (through the integral equation) with $g(y)$'s which pass through the required bars (zones). To within the limits set by our measurements these are *all* "solutions"; the ambiguity can only be removed by imposing an additional condition or criterion (not deriving from the measurements) which enables one of the set of possible $f(x)$ to be selected. We may for example ask for the smoothest $f(x)$ or the $f(x)$ with the smallest maximum deviation from the mean — but it is most important to realise that the additional condition is arbitrary. The measurements in themselves give no basis for suggesting that $f(x)$ is likely to be smooth or anything else. We select *arbitrarily* the smoothest $f(x)$ to represent all of the set of possible $f(x)$.

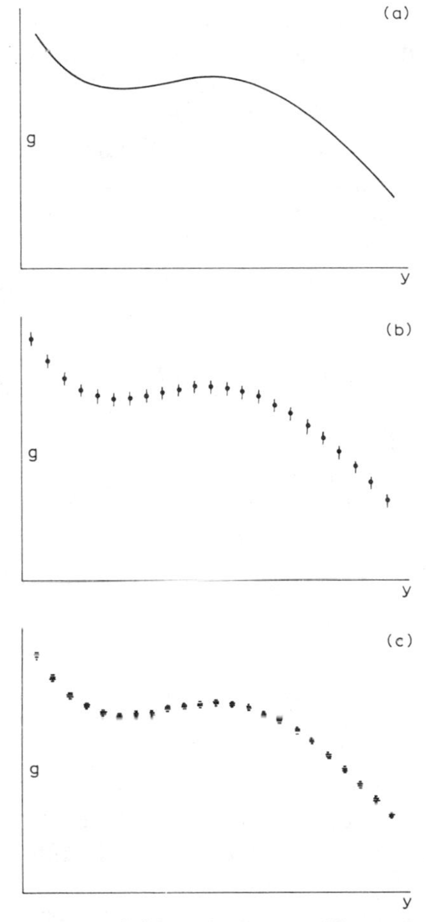

Fig. 6.1. Schematic diagram illustrating the nature of measured "functions".

Suppose $q(f)$ is some non-negative scalar measure of the deviations from smoothness in f; if f is varied until $q(f)$ becomes a minimum, the resulting f may be completely smooth in the sense that $q(f)$ will be zero. Whether a unique f is thereby selected depends on the specific nature of $q(f)$: if for $q(f)$ we use the sum of the squares of the first differences $\Sigma_{i=1}^{n-1}(f_{i+1} - f_i)^2$ evidently any constant f will give $q = 0$ and the minimum is not unique. If the expression $\Sigma_i f_i^2$ is used its minimum occurs for the unique choice

$$f = 0 \qquad \text{(i.e., } f_1 = f_2 = \ldots = 0) .$$

Now let q be incorporated with the least squares procedure so that one minimizes not $(\mathbf{A}f - g)^2$ as in the least squares method, but $|\mathbf{A}f - g|^2 + \gamma q(f)$, where γ is a parameter which can be varied from zero to infinity.

Obviously with $\gamma \to \infty$ minimization leads to $q(f) = 0$, i.e. a perfectly smooth f (judged by the measure q) but the result is the same regardless of what value $\mathbf{g}$ takes; $\gamma = 0$ on the other hand leads to the least squares solution [6.3] which as we have seen will generally exhibit large oscillations. It is apparent that intermediate values of γ produce a different kind of solution wherein we have, as it were, imposed a penalty factor for deviations from smoothness and by means of this rejected or "tuned out" the more oscillatory solutions. Since the f which minimizes $|\mathbf{A}f - \mathbf{g}|^2$ does not in general minimize $q(f)$, the solution obtained with some non-zero value for γ will give a larger value of the square norm of the residual $|\mathbf{A}f - \mathbf{g}|^2$ than is given by the least squares ($\gamma = 0$) solution.

This solution process can be looked at in the following way: a fixed value e^2 is assigned to $|\epsilon|^2 = |\mathbf{A}f - \mathbf{g}|^2$ — this is the quantity which measures the "goodness" of f as a possible solution of the equation $\mathbf{A}f = \mathbf{g}$. Of all possible vectors a small subset will have $|\mathbf{A}f - \mathbf{g}|^2 \leqslant e^2$; provided e^2 is of the same order as the uncertainties involved all vectors in the subset are acceptable as possible solutions. We now select from this subset a unique f which is the smoothest as judged by the measure $q(f)$.

Measures of smoothness

Most measures of non-smoothness are simple quadratic combinations of the f_i and are therefore quadratic forms of f which can be written $f^*\mathbf{H}f$ where $\mathbf{H}$ is usually a simple near-diagonal matrix.

Most such measures are sums of squares of individual values computed at the various i and in that case $\mathbf{H}$ is a symmetric matrix of the form $\mathbf{K}^*\mathbf{K}$, $\mathbf{K}$ being such that $\mathbf{K}f$ contains in its elements values which are squared and summed to give q.

For example, the first differences $(f_1 - f_2)$, $(f_2 - f_3)$, ..., $(f_{N-1} - f_N)$ are contained in $\mathbf{K}f$ when:

$$\mathbf{K} = \begin{bmatrix} 0 & 0 & & & & \\ 1 & -1 & 0 & \cdot & \cdot & \cdot \\ 0 & 1 & -1 & \cdot & \cdot & \cdot \\ & & \cdot & & & \\ & & & \cdot & & \\ & & & 1 & -1 & 0 \\ & & & 0 & 1 & -1 \end{bmatrix}; \quad \mathbf{K}f = \begin{bmatrix} 0 \\ f_1 - f_2 \\ f_2 - f_3 \\ \cdot \\ \cdot \\ f_{N-1} - f_N \end{bmatrix}$$

The quadratic measure $\Sigma(f_{j-1} - f_j)^2$ is therefore given by the inner product of $\mathbf{K}f$ with itself, i.e.:

$$q = (\mathbf{K}f)^* \mathbf{K}f = f^* \mathbf{K}^* \mathbf{K} f$$

So the corresponding $\mathbf{H}$ is:

$$\mathbf{H} = \mathbf{K}^* \mathbf{K}$$

Hence $\mathbf{H}$ is explicitly:

$$\mathbf{H} = \begin{bmatrix} 1 & -1 & & & & \\ -1 & 2 & -1 & & & \\ & -1 & 2 & -1 & & \\ & & & \cdot & \cdot & \\ & & & -1 & 2 & -1 \\ & & & & -1 & 1 \end{bmatrix}$$

If sums of squares of second differences were used to obtain q, the result would be:

$$\mathbf{K} = \begin{bmatrix} 0 & & & & \\ -1 & 2 & -1 & & \\ & -1 & 2 & -1 & \\ & & & \cdot & \cdot \\ & & & \cdot & \cdot \end{bmatrix}; \quad \mathbf{H} = \begin{bmatrix} 1 & -2 & 1 & 0 & \cdot & \cdot & \cdot \\ -2 & 5 & -4 & 1 & \cdot & \cdot & \cdot \\ 1 & -4 & 6 & 4 & 1 & \cdot & \cdot \\ 0 & 1 & -4 & 6 & -4 & 1 & \cdot \\ & & \cdot & \cdot & \cdot & \cdot & \cdot \end{bmatrix}$$

Another commonly used expression for q is given by the sums of the squares of the elements of f; use of this constraint selects that f which has the least value of $\Sigma f_i^2 = f^*f$; this quantity corresponds to the power in a signal. A closely related constraint is given by the variance of f, $\Sigma(f_i - \bar{f})^2$, $\bar{f}$ being the average $N^{-1} \Sigma_{i=1}^{N} f_i$. The appropriate $\mathbf{H}$ is readily obtained in each case:

(1) $q = \Sigma_i f_i^2$: $\mathbf{H}$ is evidently the identity matrix.

(2) $q = \Sigma_i (f_i - \bar{f})^2$; the differences from the average are given by:

$$\mathbf{K} = \begin{bmatrix} (1-N^{-1}) & -N^{-1} & -N^{-1} & -N^{-1} & -N^{-1} & \cdots & \\ -N^{-1} & (1-N^{-1}) & -N^{-1} & -N^{-1} & -N^{-1} & \cdots & \\ -N^{-1} & -N^{-1} & (1-N^{-1}) & -N^{-1} & -N^{-1} & \cdots & \\ \cdot & \cdot & \cdot & (1-N^{-1}) & & & \\ \cdot & \cdot & \cdot & & & & \\ -N^{-1} & -N^{-1} & \cdot & \cdot & \cdot & \cdots & (1-N^{-1}) \end{bmatrix}$$

$$= N^{-1}\|N\delta_{ij} - 1\| \qquad \left(\delta_{ij} = \begin{cases} 0, & i \neq j \\ 1, & i = j \end{cases}\right)$$

Thus the diagonal elements of H are given by:

$$h_{ii} = (N-1)\,N^{-2} + (1 - N^{-1})^2 = N^{-2}[N - 1 + (N-1)^2] = N^{-2}(N^2 - N)$$
$$= (1 - N^{-1})$$

and the off-diagonal elements are:

$$h_{ij} = (N-2)\,N^{-2} - 2(1 - N^{-1})\,N^{-1} = N^{-2}[N - 2 - 2(N-1)] = -N^{-1}$$

So **H** is equal to **K.** (This matrix therefore has the interesting property that it equals itself squared; it is in fact the sum of the identity matrix and a singular matrix containing $-N^{-1}$ in every position.)

Many other formulae may be arrived at to give other measures of smoothness, but in practice they differ very little in the final result — indeed, if they gave very different final results this would represent an ambiguity which would invalidate the whole method.

Completion of constrained solution process

Having arrived at an explicit expression for H, the solution algebra is straightforward: we minimize $q = f^*\mathbf{H}f$, subject to the constraint $|\mathbf{A}f - g|^2 \leqslant e^2$; unless a minimum of q is attained for an f lying within the subspace $|\mathbf{A}f - g| < e^2$, the inequality can be replaced by an equality and the problem becomes:

$$\left.\begin{array}{ll} \text{minimize:} & f^*\mathbf{H}f \\ \text{holding constant:} & (\mathbf{A}f - g)^*(\mathbf{A}f - g) \end{array}\right\}$$

This is a straightforward constrained extremum problem and the solution is obtained by finding an absolute (unconstrained) extremum of $(\mathbf{A}f - g)^*(\mathbf{A}f - g) + \gamma f^*\mathbf{H}f$, γ being an undetermined Lagrangian multiplier.

We therefore require that for all k:

$$\frac{\partial}{\partial f_k}\{f^*\mathbf{A}^*\mathbf{A}f - g^*\mathbf{A}f - f^*\mathbf{A}^*g + \gamma f^*\mathbf{H}f\} = 0$$

or:

$$e_k^*(\mathbf{A}^*\mathbf{A}f - \mathbf{A}^*g + \gamma\mathbf{H}f) + (f^*\mathbf{A}^*\mathbf{A} - g^*\mathbf{A} + \gamma f^*\mathbf{H})\,e_k = 0$$

The second term is the transpose of the first so if the sum vanishes the terms must vanish separately and for all k, hence:

$$(\mathbf{A}^*\mathbf{A} + \gamma\mathbf{H})f = \mathbf{A}^*g$$

or:

$$f = (\mathrm{A}^*\mathrm{A} + \gamma \mathrm{H})^{-1}\ \mathrm{A}^* g \qquad [6.4]$$

This is the equation for constrained linear inversion. The usual procedure for applying this equation is to choose several values for γ, and then post-facto decide the most appropriate value for γ by computing the residual $|\mathbf{A}\, f - g|$; if this is appreciably larger than the overall error in g due to all causes (experimental error, quadrature error, etc.) then γ is too large — the solution has been constrained too much; if $|\mathbf{A}\, f - g|$ is smaller than the estimated error in g one has an underconstrained solution — the error component or part of it has been inverted and spurious oscillations put into the solution. We shall return to the question of choosing γ after giving an example of the application of equation [6.4] to the same problem which was used earlier [$K(y, x) = x\, \mathrm{e}^{-yx}$; $f(x) = 1 + 4(x - \frac{1}{2})^2$]

6.4. SAMPLE APPLICATIONS OF CONSTRAINED LINEAR INVERSION

We have already calculated $\mathbf{A}$ and a set of values of g_i for this problem. It remains only to choose a constraint. Apropos this question it is relevant to note that the $f(x)$ selected for illustration happens to be a quadratic, so that the third differences vanish.

Of course in a practical problem one would not know this but we will show the result of applying equation [6.4] using both this constraint and the much simpler one which employs for q the quantity Σf_i^2. We have not previously computed $\mathbf{H}$ explicitly for third differences but it is easily computed. We have, for *third differences*:

$$\mathbf{K} = \begin{bmatrix} 0 & & & & & \\ 0 & 0 & & & & \\ 0 & 0 & 0 & & & \\ 1 & -3 & 3 & -1 & & \\ & 1 & -3 & 3 & -1 & \\ & & 1 & -3 & 3 & 1 \\ & & \cdot & \cdot & \cdot & \cdot \end{bmatrix}; \quad \mathbf{H} = \begin{bmatrix} 1 & -3 & 3 & -1 & & \\ -3 & 10 & -12 & 6 & -1 & \\ 3 & -12 & 19 & -15 & 6 & \\ -1 & 6 & -15 & 20 & -15 & 6 \\ & -1 & 6 & -15 & 20 & -15 \\ & & \cdot & \cdot & \cdot & \cdot \\ & & & \cdot & \cdot & \cdot \end{bmatrix}$$

For $q = \Sigma f_i^2$ we have already found $\mathbf{H} = \mathbf{I}$. When equation [6.4], with $\mathbf{H}$ selected to minimize the third differences, was applied to the set of values of g_i, it gave excellent results. That of course is not very surprising, for one is forcing the solution towards a quadratic and the original $f(x)$ from which g was computed was in fact a quadratic.

For all γ from the largest value used, which happened to be unity, down to $\gamma = 10^{-8}$, the solution agreed in every digit printed out with the original $f(x_i)$ values. The agreement, being so close, is not seen in a graphical display, so the results have been tabulated in Table 6.5. When γ reached 10^{-12} the solution was still good, but discrepancies began to appear in the fifth significant figure; the discrepancies spread to the third and occasionally the second significant figure at $\gamma = 10^{-14}$ and by $\gamma = 10^{-16}$ the solution is very poor, taking negative values in some cases. In the tables of results (Tables 6.5 and 6.6) digits in italics are those which differ from the "correct answer".

In a real inversion one cannot compare a solution or solutions with the "true" solution — only comparisons in the domain of $\mathbf{g}$ can be made. In Fig. 6.2 the "residual" $|\mathbf{g} - \mathbf{g}'|$ — expressed as a fraction of $|\mathbf{g}|$ and divided by $\sqrt{N}$ so that it becomes an r.m.s. relative residual — has been plotted against γ.

It remains less than one-thousandth of a percent down to $\gamma = 10^{-12}$ but commences to increase rapidly below that value, exceeding 100% at $\gamma = 10^{-18}$. This behavior represents the results of essentially inverting small

TABLE 6.5

Comparison of original $f(x)$ with solutions for different γ — third difference constraint

i	x_i	$f(x_i)$	f'_i (solution)				
			$\gamma = 1$	$=10^{-10}$	$=10^{-12}$	$=10^{-14}$	$=10^{-16}$
1	0.025	1.9025	1.9025	1.9025	1.902*4*	1.90*70*	1.*7344*
2	0.075	1.7225	1.7225	1.7225	1.7225	1.7*166*	1.*8750*
3	0.125	1.5625	1.5625	1.5625	1.5625	1.56*11*	1.*7656*
4	0.175	1.4225	1.4225	1.4225	1.4225	1.4*312*	*0.9688*
5	0.225	1.3025	1.3025	1.3025	1.302*4*	1.30*94*	*0.7813*
6	0.275	1.2025	1.2025	1.2025	1.202*4*	1.*1931*	1.*3438*
7	0.325	1.1225	1.1225	1.1225	1.1225	1.1*157*	*2.1250*
8	0.375	1.0625	1.0625	1.0625	1.062*7*	1.05*77*	1.*2656*
9	0.425	1.0225	1.0225	1.0225	1.022*6*	1.0*322*	*−0.6563*
10	0.475	1.0025	1.0025	1.0025	1.002*4*	1.0*127*	*−0.3125*
11	0.525	1.0025	1.0025	1.0025	1.002*3*	1.0*143*	*2.0000*
12	0.575	1.0225	1.0225	1.0225	1.022*2*	1.00*10*	*2.4063*
13	0.625	1.0625	1.0625	1.0625	1.062*4*	1.0*352*	*0.7501*
14	0.675	1.1225	1.1225	1.1225	1.122*7*	1.1*289*	*−0.1875*
15	0.725	1.2025	1.2025	1.2025	1.203*1*	1.2*295*	1.*1875*
16	0.775	1.3025	1.3025	1.3025	1.302*9*	1.3*252*	1.*4375*
17	0.825	1.4225	1.4225	1.4225	1.422*3*	1.4*116*	*0.875*
18	0.875	1.5625	1.5625	1.5625	1.562*0*	1.5*488*	1.*6875*
19	0.925	1.7225	1.7225	1.7225	1.722*1*	1.7*158*	*2.4998*
20	0.975	1.9025	1.9025	1.9025	1.902*8*	1.90*92*	1.*6875*

Digits in the solution which differ from the (known) correct answer have been italicized.

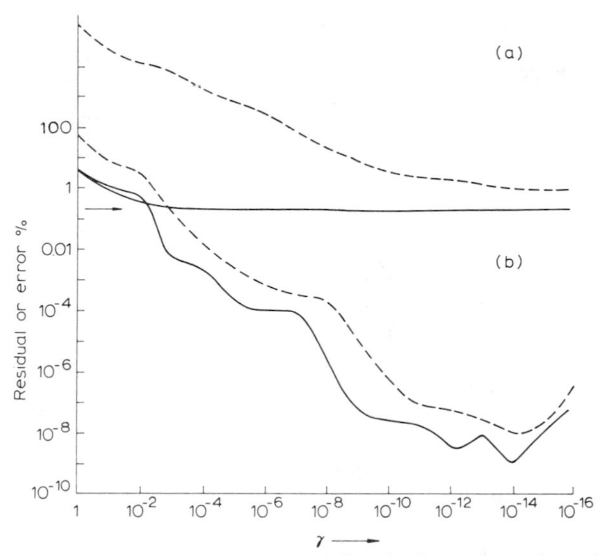

Fig. 6.2. Variation of residual for the inversion of the "standard problem" by constrained linear inversion (equation [6.4]), using the constraint of third-difference minimization. Solid curves relate to the residual $|g - g'|$; broken curves give the difference error in $f(x)$, $|f - f'|$. The lower pair contained no errors other than computer roundoff; additional error (r.m.s. value indicated by arrow) was introduced in $\boldsymbol{g}$ in the case of the upper pair of curves.

errors (roundoff and quadrature) for of course when γ becomes small enough the solution tends to the least square solution which we have seen to be oscillatory and unstable.

Effect of errors

If now error is deliberately introduced into $\boldsymbol{g}$, we obtain the solutions tabulated in Table 6.6. The error introduced was random (but not normal) with an r.m.s. value of 0.288%. We now found for larger γ acceptable solutions with (naturally) some errors, but the solution developed a mild spurious waviness at $\gamma = 10^{-6}$ and this became quite severe at 10^{-8}. While the solution was, in essence, becoming worse, the residual (Fig. 6.2a) decreased, although very slowly — the algebra being, as it were, unable to invert most of the random error and therefore making very little headway towards reducing the residual, which bottomed out around the error level 0.288% and diminished only from 0.2845% at $\gamma = 1$ to 0.2807% at $\gamma = 10^{-10}$; thereafter the residuals drifted upwards, first slowly, but more quickly from 10^{-16} down. In the meantime the solutions became larger and more oscillatory, with most elements exceeding 10^3 at $\gamma = 10^{-16}$ and changing sign eight times between $i = 1$ and $i = 20$.

TABLE 6.6

Effect of ±0.5% random error (r.m.s. error 0.288%)

i	x_i	$f(x_i)$	$\gamma = 1$	$=10^{-2}$	$=10^{-6}$	$=10^{-8}$	$=10^{-10}$
1	0.025	1.903	1.856	1.868	1.856	2.321	−3.071
2	0.075	1.723	1.688	1.694	1.721	1.632	4.824
3	0.125	1.563	1.539	1.540	1.590	1.247	5.884
4	0.175	1.423	1.408	1.406	1.463	1.160	1.055
5	0.225	1.303	1.296	1.292	1.337	1.302	−4.174
6	0.275	1.203	1.203	1.198	1.211	1.530	−5.061
7	0.325	1.123	1.129	1.124	1.090	1.663	−0.598
8	0.375	1.063	1.074	1.070	0.984	1.560	6.253
9	0.425	1.033	1.037	1.035	0.909	1.194	10.847
10	0.475	1.003	1.019	1.019	0.880	0.668	9.875
11	0.525	1.003	1.020	1.022	0.907	0.190	3.322
12	0.575	1.023	1.040	1.044	0.990	−0.016	−5.549
13	0.625	1.063	1.078	1.083	1.120	0.199	−11.939
14	0.675	1.123	1.135	1.141	1.278	0.836	−12.029
15	0.725	1.203	1.211	1.217	1.437	1.732	−5.022
16	0.775	1.303	1.306	1.311	1.569	2.590	6.191
17	0.825	1.423	1.420	1.422	1.649	3.061	15.996
18	0.875	1.563	1.552	1.551	1.627	2.836	18.149
19	0.925	1.723	1.703	1.698	1.583	1.714	7.962
20	0.975	1.903	1.873	1.862	1.426	−0.332	−15.244

Digits in the solution which differ from the (known) correct answer have been italicized.

Systematic error

When a systematic error, varying linearly with x, was introduced, the residuals behaved somewhat differently. In this case the inversion of the error, or a large component thereof, was more feasible and we find that the residual decreased to well below the r.m.s. value of the error component before it commenced to increase again, but the solutions became very poor and in fact the systematic error had a much worse effect on the solutions than did the random error, simply because the algebra was "able" to invert a sizeable part of the systematic error.

These results, although obtained for a somewhat idealized situation show some qualitative features which seem to hold quite generally:

(1) As γ is decreased the residual decreases, but for very small γ the residuals may increase again because of gross error magnification.

(2) The difference between the solution f' and the true f, $|f' - f|$ may commence to increase appreciably before the residual $|\mathbf{g}' - \mathbf{g}|$ commences to do so. In any event, the minimum of $|f' - f|$ generally does not occur in unison with that of $|\mathbf{g}' - \mathbf{g}|$, except by coincidence.

(3) In the presence of random error the residual, plotted against γ, may

become almost horizontal for several decades. This kind of behavior (see Fig. 6.2) is capable of "signalling" the magnitude of the error component, but one cannot be sure that it will always occur. Its occurrence depends on the error vector being almost entirely outside the subspace of $\boldsymbol{g}$ vectors. It does not always occur, especially if there is a systematic error component.

Turning now to the constraint $\mathbf{H} = \mathbf{I}$ which forces the solution towards the x-axis, minimizing the "moment of inertia" of the solution about that axis, we find that the inversion process was quite successful as judged by the residuals $|\boldsymbol{g}' - \boldsymbol{g}|$. The relative residual has been plotted in Fig. 6.4 for the case of no errors and the same $\pm 0.5\%$ (r.m.s. 0.288%) error used above. We find for values of γ around 10^{-5} relative residuals less than 1% and in the case of the deliberately introduced random error, its magnitude was again quite accurately signalled by the residual behavior, which bottomed out approaching 0.28%.

The solutions with or without error (Fig. 6.3) are not too good judged in comparison with $f(x)$. But if we did not know $f(x)$, we would have only the residual $|\boldsymbol{g}' - \boldsymbol{g}|$ to look at (Fig. 6.4) and the solutions are just as good by that criterion as the "excellent" solutions obtained with the third difference constraint. Quite in keeping with the minimum "moment of inertia" constraint, the larger departures from the x-axis near $x = 0$ and $x = 1$ have been severely curtailed and smoothed, but this has been done in a way which maintains the values of $\int K_i(x)\, f'(x)\, dx$ very close to those of $\int K_i(x)\, f(x)\, dx$. The solutions of Fig. 6.3 are typical of those obtained with constraints which conflict with the original $f(x)$ — in the sense that $f(x)$ does not minimize the relevant quadratic form. The solution tends to follow the original $f(x)$ but becomes more "wavy". One of the prime difficulties in inversion

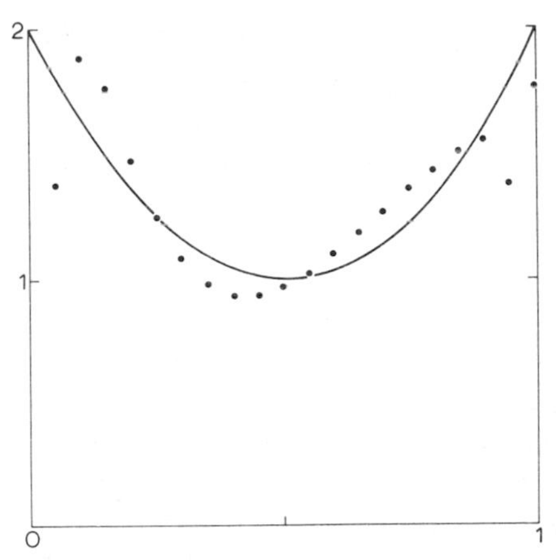

Fig. 6.3. The solution obtained to the "standard problem" using the constraint given by $\mathbf{H} = \mathbf{I}$.

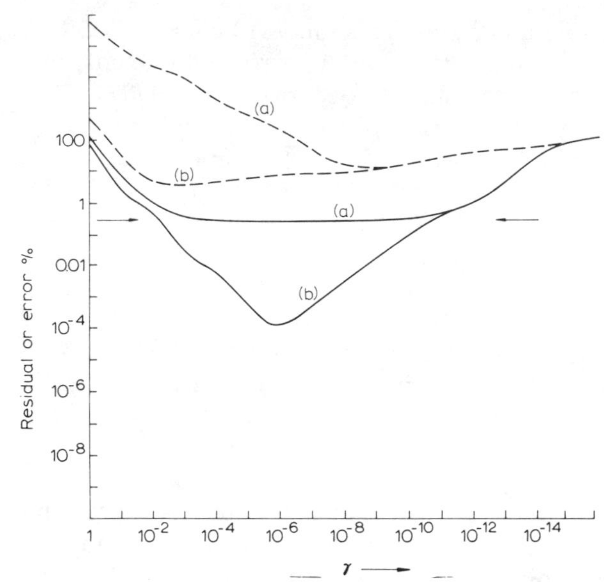

Fig. 6.4. Variation of residuals in the case of the solution with H = I. The curves have the same significant as those of Fig. 6.2; error added to g in the case of curves (a), r.m.s. value of error indicated by arrows.

work is to establish the probable reality or otherwise of such wave-like fluctuations.

Having obtained the solutions with H = I it would clearly be unwise to accept the solution for $\gamma = 10^{-5}$ or 10^{-6}, although if all solutions are equally probable and acceptable that is all that can be done. However, having accepted the postulate that a smoothest acceptable solution should be sought it is desirable in a case like this to look at the solutions first obtained (i.e. Fig. 6.3) and consider the desirability of modifying the constraint. These solutions show a broad minimum around $x = 0.5$ and move upward to larger values as x goes towards 0 and 1 — a kind of behavior which clearly does not minimize Σf_i^2. Having recognized this "desire" in the solution, one can legitimately alter the definition of smoothness. For example, we can:

(1) Use another quite independent measure of smoothness, e.g. Σ(second differences)^2.

(2) Weigh the solution according to the solution just obtained and solve the problem anew, i.e. take a smoothed version of the first solution as $f_0(x)$ and solve the problem $\boldsymbol{g} = \int \boldsymbol{k}(x)\, f_0(x)\, u(x)\, dx$ for $u(x)$ using the constraint H = I (all that this necessitates is replacement of I by a diagonal matrix with elements *inversely* proportional to the corresponding value of $f_0(x)$).

(3) Rephrase the smoothness measure in any other way to take into account what the first solutions indicated about the behavior of $f(x)$.

In many real physical inversions there are known a priori constraints on $f(x)$ and in many situations after a first attempt at inversion has given solutions that are mathematically acceptable but physically implausible it is realized that the physics of the situation biases $f(x)$ towards a certain kind of behavior.

It only requires the incorporation of these expectations into the constraints to push the solution in the same direction. In some cases a weighting of the elements of $f(x)$ in q is appropriate, in other cases it may be useful to divide out an average or expected $f_0(x)$ as was mentioned above. The other option, especially appropriate when a statistically averaged value is known, is to measure the departure of $f(x)$ not from the x-axis, but from the statistical average. We will give the appropriate algebraic formulae later in this chapter; they are all very simple modifications of equation [6.4].

It is to be noted that the residuals referred to just now are the norms of the differences between the original $\mathbf{g}$ and $\mathbf{g}'$, which is obtained when the solution is inserted into the integral $\int \mathbf{k}(x)\, f'(x)\, dx$ or the quadrature expression $\mathbf{A}\mathbf{f}'$. These are the only residuals available to test the inversion when the true solution is unknown; in the examples above, we know the true solution and could evaluate residuals to measure $|\mathbf{f}' - \mathbf{f}|$. It is important to notice that a reduction in $|\mathbf{g}' - \mathbf{g}|$ does *not* imply a reduction in $|\mathbf{f}' - \mathbf{f}|$. In fact when the solutions become unstable $|\mathbf{g}' - \mathbf{g}|$ usually continues to diminish but $|\mathbf{f}' - \mathbf{f}|$ increases very rapidly. Some of the results may show an increase in $|\mathbf{g}' - \mathbf{g}|$ when γ is reduced beyond a certain stage, but this is caused by numerical errors which are aggravated when the inverse matrix elements become enormously large. We cannot use this behavior to gauge the usefulness or accuracy of the solution — generally speaking one can be sure that γ is too small when $|\mathbf{g}' - \mathbf{g}|$ commences to increase again, but $|\mathbf{f}' - \mathbf{f}|$ may have commenced to increase at a much earlier stage in the proceedings. What must be done is to make a realistic but safe upper estimate of the error component and select a value for γ which is on the side of the curve of $|\mathbf{g}' - \mathbf{g}|$ versus γ where $|\mathbf{g}' - \mathbf{g}|$ decreases with decreasing γ but is still not less than the estimated error. If, for example, the following residuals were obtained:

γ	$\lvert g' - g\rvert / \lvert g\rvert$
10^{-4}	4%
10^{-5}	2.5%
10^{-6}	1.5%
10^{-7}	0.3%
10^{-8}	0.05%
10^{-9}	0.008%
10^{-10}	0.15%
10^{-11}	10%

Then if the combined errors are confidently believed to be 1% or less, the solution for $\gamma = 10^{-6}$ can be accepted; if we are not very certain about the figure of 1%, it would be better to use $\gamma = 10^{-5}$ (for if we overestimate γ we obtain a useful but oversmoothed solution; too small a value of γ can give large instability effects).

The constrained inverstion formula as a least squares result

As we have already seen a set of M linear relations between N unknowns can be solved in a least squares sense, the resulting solution being one which minimizes the sum of the squares of the residuals — if $M > N$ the least squares solution can still be obtained but none of the linear relationships may then be *exactly* satisfied by the solution.

If certain members of the set of linear relations are to be given more weight than others, this is very easily taken care of. In the system:

$$\left.\begin{array}{l} a_{11}f_1 + a_{12}f_2 + \ldots + a_{1N}f_N = g_1 \\ a_{21}f_1 + a_{22}f_2 + \ldots + a_{2N}f_N = g_2 \\ \cdot \\ \cdot \\ a_{M1}f_1 + a_{M2}f_2 + \ldots + a_{MN}f_N = g_N \end{array}\right\}$$

weight w_1 may be assigned to the first relations, w_2 to the second and so on, simply by including the w's in the minimization procedure. If r_i denotes the ith residual $g_i - \Sigma_j\, a_{ij}\, f_j$, the weighting is achieved simply by minimizing $\Sigma_i\, w_i\, r_i^2$ rather than Σr_i^2. One obtains a solution of the same form as equation [6.3] except that in place of $\mathbf{A}^*\mathbf{A}$, i.e. $\|\Sigma_k\, a_{ki}\, a_{kj}\|$, one has $\|\Sigma_k\, w_k\, a_{ki}\, a_{kj}\|$, and in place of $\mathbf{A}^*\boldsymbol{g} = \|\Sigma_k\, a_{ki}\, g_k\|$ there occurs $\|\Sigma_k\, w_k\, a_{ki}\, g_k\|$ — in other words, the various vector inner products are replaced by weighted sums of products of elements, a kind of weighted inner product.

Consider now a system of linear equations:

$$\left.\begin{array}{l} a_{11}f_1 + a_{12}f_2 + \ldots + a_{1N}f_N = g_1 \\ a_{21}f_1 + a_{22}f_2 + \ldots + a_{2N}f_N = g_2 \\ \cdot \\ \cdot \\ a_{M1}f_1 + a_{M2}f_2 + \ldots + a_{MN}f_N = g_M \end{array}\right\}$$

and a system of constraining equations such as:

$$\left.\begin{array}{lllll} f_1 & & & & = 0 \\ & f_2 & & & = 0 \\ & & \cdot & & \\ & & & \cdot & \\ & & & & f_N = 0 \end{array}\right\}$$

In a sense these are contradictory: quite obviously $f_1 = f_2 = f_3 = \ldots, = f_N = 0$ satisfies the N constraint equations exactly, but not in general the first M equations involving the a_{ij}'s. Suppose we assign a weight γ to each of the constraint equations and a weight of unity to the first set of equations, and find a least squares solution. The weighted inner products of columns i and j obtain no contributions from the constraint equations and are simply $\Sigma a_{ki}\, a_{kj}$ for $i \neq j$, as in the ordinary unweighted least squares formula. The weighted inner products of columns with themselves give an additional contribution of γ in each i, i element. Since the right-hand side is zero in the case of the constraint equations they make no contribution to the right-hand side of the final equation. Thus the solution to the weighted least squares problem is:

$$f = (\mathbf{A}^* \mathbf{A} + \gamma \mathbf{I})^{-1}\, \mathbf{A}^* g$$

which is the constrained inversion formula [6.4] for the case $\mathbf{H} = \mathbf{I}$.

Other constrained inversions can similarly be derived as weighted least squares problems, where the uncertainty associated with the direct solution is taken care of by including (but with lesser weight) suitable additional relations which by themselves would produce a completely smooth f. For example, the second difference constraint obtained by using for $\mathbf{H}$ the form already derived can also be derived by including, each with weight γ, the additional relations:

$$\left.\begin{array}{l}
-1 \cdot f_1 + 2 \cdot f_2 - 1 \cdot f_3 = 0 \\
\quad -1 \cdot f_2 + 2 \cdot f_3 - 1 \cdot f_4 = 0 \\
\quad\quad -1 \cdot f_3 + 2 \cdot f_4 - 1 \cdot f_5 = 0 \\
\quad\quad\quad -1 \cdot f_4 + 2 \cdot f_5 - 1 \cdot f_6 = 0 \\
\quad\quad\quad\quad \cdot \\
\quad\quad\quad\quad \cdot \\
\quad\quad\quad\quad -1 \cdot f_{N-2} + 2 \cdot f_{N-1} - 1 \cdot f_N = 0
\end{array}\right\}$$

The diagonal elements, away from the corner are incremented by $(1^2 + 2^2 + 1^2)\,\gamma = 6\gamma$, the adjacent elements above and below the diagonal by -4γ, and their neighbors by $+\gamma$. The right-hand side of the final equation is unchanged from the simple least squares form, and we again obtain:

$$f = (\mathbf{A}^* \mathbf{A} + \gamma \mathbf{H})^{-1} \mathbf{A}^* g$$

with H being, as before:

$$\begin{bmatrix} 1 & -2 & 1 & & & & \\ -2 & 5 & -4 & 1 & & & \\ 1 & -4 & 6 & -4 & 1 & & \\ & 1 & -4 & 6 & -4 & 1 & \\ & \cdot & \cdot & \cdot & \cdot & \cdot & \cdot \\ & \cdot & \cdot & \cdot & \cdot & \cdot & \cdot \end{bmatrix}$$

Weighted least squares solutions are therefore in every way identical with those given by constrained linear inversion — only the approach to the problem, and in particular the specific way in which γ was introduced, are different.

The weighted least squares approach is quite helpful in formulating what might be called "tight" constraints. If, for example, external a priori considerations made it necessary that $f(x_1) = f(x_N) = 2$, so that f_1 and f_N are known, then the addition of the relations $f_1 = 2$ and $f_N = 2$ with a large enough weight β produce solutions of the required character. This is achieved simply by adding β to the first and last diagonal element of the matrix $(\mathbf{A}^*\mathbf{A} + \gamma \mathbf{H})$ and 2β to the first and last elements of the righthand side vector $\mathbf{A}^*\boldsymbol{g}$. Applying this constraint in the case of the example already used several times, a marked improvement was achieved, as Table 6.7 shows.

Forms of constraint when the statistics of the unknown distribution are known

In some inversion problems, nothing at all is known about the unknown function $f(x)$. One can obtain an inversion using some objective constraint such as minimization of variance, but the solution thus obtained has a considerable element of ambiguity about it. One can have some confidence about the general shape of the solution but finer structure is another question. In other situations a considerable amount of background exists — perhaps measurements by direct methods have been made, as is the case with atmospheric temperature profiles, which were measured by balloon-borne and aircraft-borne instruments long before satellite-borne indirect measurements were possible. These measurements showed, for example, that relatively steady gradients were to be expected through certain layers of the atmosphere while sharper excursions and reversals of gradient tended to occur at more or less predictable levels. Furthermore fluctuations from the norm at different levels tended to be correlated with each other, in the sense that some certain layer being warmer than average might imply a high probability of certain other layers being colder than average, and so on. It is possible

TABLE 6.7

Inversion of standard problem with $f(0)$ and $f(1)$ tightly constrained to the value 2.0 ($\beta = 1$, $\gamma = 10^{-5}$, second-difference constraint)

x	Inversion for $f(x)$	True $f(x)$
0.025	1.903	1.9025
0.075	1.725	1.7025
0.125	1.554	1.5625
0.175	1.399	1.4225
0.225	1.268	1.3025
0.275	1.168	1.2025
0.325	1.099	1.1225
0.375	1.060	1.0625
0.425	1.043	1.0225
0.475	1.044	1.0025
0.525	1.056	1.0025
0.575	1.077	1.0225
0.625	1.105	1.0625
0.675	1.143	1.1225
0.725	1.198	1.2025
0.775	1.276	1.3025
0.825	1.384	1.4225
0.875	1.527	1.5625
0.925	1.704	1.7225
0.975	1.903	1.9025

As usual digits in the inversion solution have been italicized when it differs from the corresponding digit in the true solution.

to construct constraints which allow for such correlations — but it is important to realise that by so doing one is pushing the indirectly sensed solutions towards conformity with the body of past data obtained by more direct methods. Such a forcing is difficult to justify if it is applied, for example, in regions of the globe where no body of past data exist.

A straightforward method of taking into account the tendencies existing in a body of past data is to construct from the past data a suitable set of base functions for approximating the unknown $f(x)$. In a crude way this can be done by deriving the mean $\overline{f(x)}$ of all past data and finding a constrained solution which minimizes the mean-square departure from this mean.

To do this one merely uses for the quadratic form q the square norm of $(\boldsymbol{f} - \boldsymbol{b})$, $\boldsymbol{b}$ being the expected (mean) value towards which we wish to bias $\boldsymbol{f}$. In the usual way, the solution is obtained by finding the extremum of:

$$(\mathbf{A}\boldsymbol{f} - \boldsymbol{g})^*(\mathbf{A}\boldsymbol{f} - \boldsymbol{g}) + \gamma(\boldsymbol{f} - \boldsymbol{b})^*(\boldsymbol{f} - \boldsymbol{b})$$

which is given by:

$$e_k^* A^* A f - e_k^* A^* g + \gamma e_k^* f - \gamma e_k^* h = 0 \quad (k = 1, 2, \dots N)$$

or:

$$f = (A^* A + \gamma I)^{-1} (A^* g + \gamma h) \qquad [6.5]$$

When there is a reasonable basis for choosing a particular h, this formula gives a useful improvement. But if h is unrealistic the attempt to force f to stay close to h results in spurious oscillations of the solution. An extreme example of this is shown in Fig. 6.5 where h was deliberately chosen to be quite different from the f which had been used to generate the set of g_i values used for the inversion. It is noteworthy that the solution f has not been pushed towards h in toto (which intuitively might have been expected). This kind of behavior is frequently produced by the application of a "contradictory" constraint and it compounds the difficulty of sorting out the true and the false in structural features, especially when no background information exists about the statistics of $f(x)$.

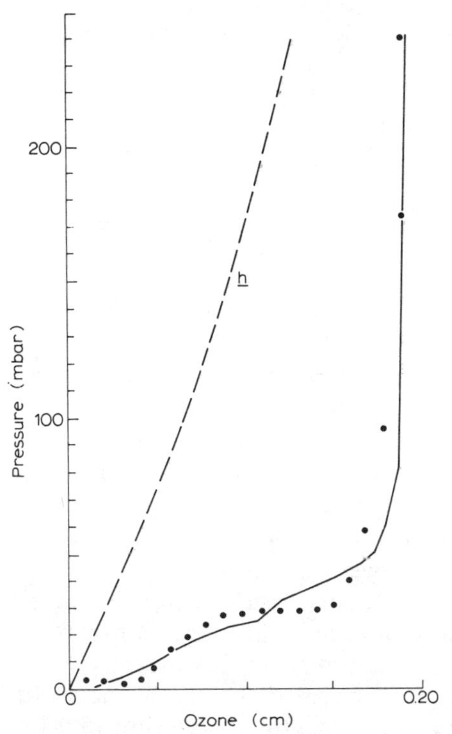

Fig. 6.5. An inversion related to an ozone-sensing problem, in which the "bias" vector h was deliberately made very different from any possible form which the physics would allow. The solution responded by developing an undulation around the solution obtained with a realistic choice of bias. The circles give the distorted solution, the solid line is the solution obtained with a realistic bias vector.

Use of characteristic patterns (orthogonal empirical functions)

When the statistics are extensive and where there are good grounds for believing that they are applicable to the solution being sought, it is possible to refine the solution process beyond the mean behavior demanded by equation [6.5]. If one possesses a large set of values taken by a function $f(x)$ in the past, then if that set is $f_1(x)$, $f_2(x)$, ..., $f_N(x)$ one could obtain a reasonable basis for approximating any future $f(x)$ by using a linear combination of the set, i.e., by setting:

$$f(x) = \sum_j c_j f_j(x)$$

and solving for the c_j's.

This, however, may involve a very large number of coefficients and is clearly not efficient. The same effect can be produced by producing a set of orthogonal functions $\phi_1(x)$, $\phi_2(x)$, etc., by forming linear combinations of the N functions of the set $f_i(x)$, $i = 1, 2, ..., N$. Since many members of the set will be almost identical with other members of the set there is little likelihood that a full set of N orthogonal functions will have to be employed — there will be members of the set of f's which differ by less than the errors of measurement from other members, or from linear combinations of a few other members.

The efficiency of any set of orthogonal functions for representing the functions from which they were constructed can be examined most readily if we consider the residual or error term remaining when the series is truncated. If we construct the full set of N orthogonal functions $\hat{\phi}_1(x)$, $\hat{\phi}_2(x)$, ..., $\hat{\phi}_N(x)$ then every $f_i(x)$ is exactly equal to the sum of the series $\Sigma_k\ b_{ik}\ \hat{\phi}_k(x)$, b_{ik} being of course $\int f_i(x)\ \hat{\phi}_k(x)\ dx$. If any linear combinations of the $f_i(x)$ happen to vanish, some of the N orthogonal functions will be zero — less than N orthogonal functions will then be needed, but in any event N will always suffice and each of the following equations are exact:

$$f_1(x) = b_{11}\hat{\phi}_1(x) + b_{12}\hat{\phi}_2(x) \;\Big|+ \ldots + b_{1N}\hat{\phi}_N(x)$$

$$f_2(x) = b_{21}\hat{\phi}_1(x) + b_{22}\hat{\phi}_2(x) \;\Big|+ \ldots + b_{2N}\hat{\phi}_N(x)$$

$$\cdot$$

$$\cdot$$

$$f_N(x) = b_{N1}\hat{\phi}_1(x) + b_{N2}\hat{\phi}_2(x) \;\Big|+ \ldots + b_{NN}\hat{\phi}_N(x)$$

There are many ways in which $\hat{\phi}$'s can be constructed; the general relationship:

$$\hat{\phi}(x) = \mathbf{A} f(x)$$

A being an arbitrary matrix, evidently will produce orthogonal and normal

$\hat{\phi}_i$'s provided only that $\int \hat{\phi}(x)\, \hat{\phi}^*(x)\, dx = \mathbf{I}$, that is that:

$$\mathbf{ACA}^* = \mathbf{I}$$

$$(c = \int f(x) f^*(x)\, dx)\ .$$

Suppose, however, that the equations for $f_i(x)$ above are truncated after two terms, where the broken line was drawn. There will then be residuals which may be written $r_{i2}(x)$:

$$f_1(x) = b_{11}\hat{\phi}_1(x) + b_{12}\hat{\phi}_2(x) + r_{12}(x)$$

$$f_2(x) = b_{21}\hat{\phi}_1(x) + b_{22}\hat{\phi}_2(x) + r_{22}(x)$$

$$\cdot \quad \cdot \quad \cdot \quad \cdot$$

$$\cdot \quad \cdot \quad \cdot \quad \cdot$$

$$f_N(x) = b_{N1}\hat{\phi}_1(x) + b_{N2}\hat{\phi}_2(x) + r_{N2}(x)$$

The error in the two-term approximations can evidently be gauged by looking at $\Sigma|r_{i2}(x)|^2$ and comparing it with $\Sigma|f_i(x)|^2$; for an exact representation the former quantity must vanish identically, while a very poor approximation is implied by $\Sigma|r_{i2}(x)|^2 \sim \Sigma|f_i(x)|^2$

Evidently, out of all possible sets of $\{\hat{\phi}_1(x), \hat{\phi}_2(x)\}$ there is one which will minimize the residual $\Sigma_i|r_{i2}(x)|^2$ and that set is optimal from the point of view of approximating the set of N functions. Similarly there are sets of $\hat{\phi}$'s which minimize $\Sigma|r_{i1}(x)|^2$, $\Sigma|r_{i2}(x)|^2$, $\Sigma|r_{i3}(x)|^2$ and so on. If ρ_1^2, ρ_2^2, etc., represent the values attained by $\Sigma|r_{i1}(x)|^2$, $\Sigma|r_{i2}(x)|^2$, etc., when the optimal $\hat{\phi}$'s are used for each approximation, then clearly a plot of ρ_k^2 against k, or better still a plot of ρ_k^2 divided by its maximum possible value $\Sigma|f_i(x)|^2$, shows what proportion of the overall norms remain unaccounted for at any stage of approximation. Furthermore, if on such a plot (Fig. 6.6) the uncertainty in the $f_i(x)$ — caused, for example, by measurement errors — is indicated, one can immediately see how far the expansion can be taken before the residual is less than the uncertainty with which the $f_i(x)$ are themselves known. To carry an expansion beyond that number of terms is clearly pointless. The number of terms which can meaningfully be carried in the presence of error can be regarded as a modified "dimension" of the function space spanned by $f_1(x), f_2(x), ..., f_N(x)$, just as the number of terms m needed for exact representation gives the dimension of the function space in an exact, analytic sense.

Having obtained, say, the first two optimal functions $\hat{\phi}_1(x)$ and $\hat{\phi}_2(x)$, the determination of $\hat{\phi}_3(x)$ can be regarded as the determination of the *first* optimal orthogonal function for the modified functions $f_1(x) - b_{11}\ \phi_1(x) - b_{12}\ \hat{\phi}_2(x)$; $f_2(x) - b_{21}\ \hat{\phi}_1(x) - b_{22}\ \hat{\phi}_2(x)$, etc., so the problem of obtaining the set of ϕ's reduces to that of getting the first, $\hat{\phi}_1(x)$ — thereafter all that is involved is a repetition of the same procedure with functions modified by

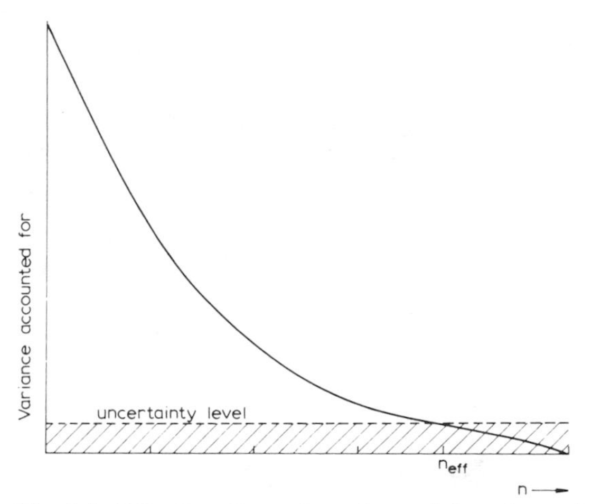

Fig. 6.6. Effective dimension of set of functions when noise (error) is present.

the removal from them of their components in the direction of those ϕ's which have already been found.

To find $\hat{\phi}_1(x)$, the residual norms $\Sigma_i |r_{i1}|^2$ must be minimized; equivalently $\Sigma_i b_{i1}^2$ must be maximized. This is accomplished by:

$$\hat{\phi}_1(x) = \xi_1 f_1(x) + \xi_2 f_2(x) + \ldots + \xi_N f_N(x)$$

if the ξ's are chosen to maximize $\Sigma_i b_{i1}^2 = \Sigma_i [\int f_i(x)\, \Sigma \xi_k\, f_k(x)\, \mathrm{d}x]^2 = \xi^* \mathbf{C}^* \mathbf{C} \xi$ where $\mathbf{C}$ is the covariance matrix $\| \int f_i(x)\, f_j(x)\, \mathrm{d}x \|$.

Thus again we return to the familiar problem of minimizing a quadratic form and the solution of course is that (apart from a scalar multiplier) ξ must be that eigenvector of $\mathbf{C}^*\mathbf{C}$ which is associated with largest eigenvalue.

If $\boldsymbol{u}$ is an eigenvector of the symmetric matrix $\mathbf{C}$ it must also be an eigenvector of $\mathbf{C}^*$ and of $\mathbf{C}^*\mathbf{C}$; in fact if $\mathbf{C}\boldsymbol{u} = \lambda \boldsymbol{u}$, then $\mathbf{C}^*\mathbf{C}\boldsymbol{u} = \lambda^2 \boldsymbol{u}$. Thus $\boldsymbol{u}$ is indeed an eigenvector of $\mathbf{C}^*\mathbf{C}$ and the maximum value of the quadratic form $\boldsymbol{u}^*\mathbf{C}^*\mathbf{C}\boldsymbol{u}$ is simply λ^2 (λ being an eigenvalue of $\mathbf{C}$ and λ^2 an eigenvalue of $\mathbf{C}^*\mathbf{C}$).

Since $\phi_1(x)$ is required to be a normalized function ξ must be chosen so that:

$$1 = \int \hat{\phi}_1(x)\, \hat{\phi}_1(x)\, \mathrm{d}x = \int \xi^* f(x)\, f^*(x)\, \xi \mathrm{d}x$$

$$= \xi^* \mathbf{C} \xi$$

Putting $\xi = \alpha \boldsymbol{u}$, the normalization condition becomes $\alpha^2 \lambda = 1$; so the normalization factor α is $\lambda^{-1/2}$ and the properly normalized optimal orthogonal function is given by:

$$\hat{\phi}_1(x) = \lambda_1^{-1/2} \boldsymbol{f}^*(x) \boldsymbol{u}_1 = \lambda_1^{-1/2} \sum_k u_{k1} f_k(x)$$

λ_1 being the largest eigenvalue of $\mathbf{C}$ and $\boldsymbol{u}_1$ the corresponding eigenvector $(u_{11}, u_{21}, \ldots, u_{N1})$.

The sum of the norms $\Sigma |f_i(x)|^2$ is evidently equal to the trace of $\mathbf{C}$ and therefore to the sum of the eigenvalues of $\mathbf{C}$; out of this the amount accounted for in the first terms of the expansions is $\Sigma_i\ b_{i1}^2$. This is readily evaluated, since:

$$\boldsymbol{b}_1 = \begin{bmatrix} b_{11} \\ b_{21} \\ \cdot \\ \cdot \\ b_{N1} \end{bmatrix} = \int \begin{bmatrix} f_1(x) \\ f_2(x) \\ \cdot \\ \cdot \\ f_N(x) \end{bmatrix} \hat{\phi}_1(x)\,\mathrm{d}x$$

$$= \int \boldsymbol{f}(x) \lambda_1^{-\frac{1}{2}} \boldsymbol{f}^*(x)\,\boldsymbol{u}_1\,\mathrm{d}x$$

$$= \lambda_1^{-1/2} \mathbf{C} \boldsymbol{u}_1 = \lambda_1^{1/2} \boldsymbol{u}_1$$

so that $\Sigma b_{i1}^2 = \boldsymbol{b}_1^* \boldsymbol{b}_1 = \lambda_1$

Thus the proportion of the overall sum of norms which is accounted for when one term of the expansions is taken is given by $\lambda_1 / \Sigma \lambda_k$.

To obtain the second optimal function $\phi_2(x)$ the procedure is the same but in place of the covariance matrix $\mathbf{C}$ we use the covariance matrix given by the modified functions $f_1(x) - b_{11}\ \hat{\phi}_1(x)$, $f_2(x) - b_{21}\ \hat{\phi}_1(x)$, etc. In terms of the vectors $\boldsymbol{f}(x)$, $\boldsymbol{b}_1$, the covariance matrix now becomes:

$$\int \{\boldsymbol{f}(x) - \boldsymbol{b}_1 \hat{\phi}_1(x)\} \{\boldsymbol{f}(x) - \boldsymbol{b}_1 \hat{\phi}_1(x)\}^* \,\mathrm{d}x$$

noting that $\boldsymbol{b}_1 = \int \hat{\phi}_1(x)\ \boldsymbol{f}(x)\ \mathrm{d}x$ and that $\int \hat{\phi}_1(x)\ \hat{\phi}_1(x)\ \mathrm{d}x = 1$, this is readily found to be:

$$\mathbf{C} - \boldsymbol{b}_1 \boldsymbol{b}_1^* = \mathbf{C} - \lambda_1 \boldsymbol{u}_1 \boldsymbol{u}_1^*$$

It is easily verified that this matrix possesses the same eigenvalues and eigenvectors as $\mathbf{C}$, except that the eigenvalue associated with $\boldsymbol{u}_1$ has been reduced to zero. In fact:

$$(\mathbf{C} - \lambda_1 \boldsymbol{u}_1 \boldsymbol{u}_1^*) \boldsymbol{u}_k = \mathbf{C} \boldsymbol{u}_k = \lambda_k \boldsymbol{u}_k \qquad (k \neq 1)$$

while $(\mathbf{C} - \lambda_1 \boldsymbol{u}_1 \boldsymbol{u}_1^*) \boldsymbol{u}_1 = \mathbf{C} \boldsymbol{u}_1 - \lambda_1 \boldsymbol{u}_1 = 0$

Evidently, therefore, the second optimal function $\hat{\phi}_2(x)$ is given by $\lambda_2^{-1/2}$-$\boldsymbol{u}_2^*\ \boldsymbol{f}(x)$; of the sum $\Sigma_i\ f_i(x)^2$ the second terms account for λ_2. Similar results hold throughout the sequence of partial sums, with:

$$\hat{\phi}_k(x) = \lambda_k^{-1/2} \boldsymbol{u}_k^* \boldsymbol{f}(x) = \lambda_k^{-1/2} \sum_j u_{jk} f_j(x)$$

$\sum_{1}^{k} \lambda_j \Big/ \sum_{1}^{N} \lambda_j$ = proportion of overall variation accounted for by the sum to k terms

These optimal approximating functions are found in many fields, under several different names. The names "principal components", "orthogonal empirical functions" and "characteristic patterns" are frequently encountered.

Application to inversion

To apply these results in inversion problems the most direct method is to expand the unknown $f(x)$ in terms of the optimal set of $\hat{\phi}$'s:

$$f(x) = \beta_1 \hat{\phi}_1(x) + \beta_2 \hat{\phi}_2(x) + \ldots + \beta_l \hat{\phi}_l(x)$$

taking only as many terms as are needed to approximate the ensemble of distributions $f_1(x), f_2(x), \ldots, f_n(x)$ to within the probable error in the $f_i(x)$. One then solves for the unknowns $\beta_1, \beta_2, \ldots, \beta_l$, introducing the statistical background by noting that λ_1 represents the average value of β_1^2 taken over the ensemble $f_1(x), f_2(x), \ldots, f_n(x)$, λ_2 the average of β_2^2 and so on. (This follows directly from the way in which the optimal set of $\hat{\phi}$'s was produced.)

Averaged over the statistical ensemble (e.g. the past data) one would find that the coefficients β_1, β_2, etc., decreased, whereas $\lambda^{-1/2}\ \beta_1, \lambda_2^{-1/2}\ \beta_2$, etc., would each give a mean square value of unity. Thus in terms of the unknown β's it is appropriate on the basis of the statistics to ask that the quadratic form $\Sigma\lambda_k^{-1}\ \beta_k^2 = \beta^* \Lambda^{-1}\ \beta$ be adopted to produce a constrained solution, i.e. $|\mathbf{A}\boldsymbol{f} - \boldsymbol{g}|^2$ is assigned a suitable value based on the errors present, and out of the set of $\boldsymbol{f}$ thereby admitted the most probable solution is taken to be that which minimizes the quadratic $\beta^* \Lambda^{-1}\ \beta$. This can be done in several ways, different in detail but not in essence.

If the ϕ's are known and tabulated, the problem is most directly solved by obtaining an equation of the form:

$$\boldsymbol{g} = \mathbf{B}\,\boldsymbol{\beta}$$

and solving for $\boldsymbol{\beta}$, using the constraint $\boldsymbol{\beta}^*\ \mathbf{\Lambda}^{-1}\boldsymbol{\beta}$ = minimum. This gives:

$$\boldsymbol{\beta} = (\mathbf{B}^*\mathbf{B} + \gamma\,\mathbf{\Lambda}^{-1})^{-1}\ \mathbf{B}^*\boldsymbol{g} \qquad [6.6]$$

The matrix $\mathbf{B}$ is derived readily since:

$$\boldsymbol{g} = \int \boldsymbol{k}(x)\, f(x)\, \mathrm{d}x$$

$$= \int \boldsymbol{k}(x)\, \hat{\boldsymbol{\phi}}^*(x)\, \mathrm{d}x\, \boldsymbol{\beta}$$

Hence:

$$B = \int k(x)\, \hat{\phi}^*(x)\, dx = \| \int k_i(x)\, \hat{\phi}_j(x)\, dx \|$$

(i.e. dot product of ith kernel with jth orthogonal function $= b_{ij}$). The equation derived for β can be expressed also as an explicit equation for the solution $f'(x)$ — we will use the prime sign to emphasize that $f'(x)$ is the scalar-valued solution while $\boldsymbol{f}(x)$ is the array of past values of $f(x)$:

$$f'(x) = \hat{\phi}^*(x)\, \beta$$
$$= \boldsymbol{f}^*(x)\, U \Lambda^{-1/2}(B^*B + \gamma \Lambda^{-1})^{-1}\, B^*\boldsymbol{g}$$

This form, however, is not particularly useful, especially when there are a large number of functions in the vector $\boldsymbol{f}(x)$; in some situations this number may run to thousands and the computaton involved is unnecessarily tedious. It is better to determine first the number of $\hat{\phi}$'s which are needed to reconstruct any of the set $f_i(x)$ to better than the uncertainty with which the $f_i(x)$ are known, and thereafter to use only those $\hat{\phi}$'s. In most practical examples the number of $\hat{\phi}$'s is rarely more than ten even when the number of $f_i(x)$ and the number of tabular values of x are both quite large. (The reason of course being that the $f_i(x)$ are far from mutually orthogonal and the value of any $f_i(x)$ at some particular value of x is highly correlated with the values which it takes elsewhere.)

6.5. ALGEBRAIC NATURE OF CONSTRAINED LINEAR INVERSION

It has been emphasized more than once in the preceding chapters that the fundamental cause of instabilities in inversions is the presence of one or several near-vanishing linear combinations of the kernels, a condition which we have shown to produce corresponding near-vanishing eigenvalues in the covariance matrix $\|\int K_i(x)\, K_j(x)\, dx\|$. It is intuitively obvious that the quadrature matrix $\mathbf{A}$ will similarly be affected and it can easily be demonstrated by noting that the existence of a very small combination $\Sigma_j\, \xi_j\, K_j(x)$ implies that for an arbitrary function $u(x)$, the quantity $\Sigma \xi_j \int K_j(x)\, u(x)\, dx$ must be small and hence the quadrature approximation thereto, $\boldsymbol{\xi}^* \mathbf{A}\boldsymbol{u}$, also must be small ($\boldsymbol{u}$ being the vector with elements $u(x_1)$, $u(x_2)$, ..., $u(x_N)$); but $u(x)$, being arbitrary, can be chosen so that $\boldsymbol{u} = \boldsymbol{\xi}$ and therefore $\boldsymbol{\xi}^*\mathbf{A}\boldsymbol{\xi}$ is small and the quadrature matrix $\mathbf{A}$ must also have at least one eigenvalue less than or equal to $\boldsymbol{\xi}^*\mathbf{A}\boldsymbol{\xi}$. $\mathbf{A}^*\mathbf{A}$ similarly must have small eigenvalues.

It is fairly clear why the constrained inversion formula is stabilized by the additional $\mathbf{H}$ term and it is especially obvious when $\mathbf{H}$ is the identity matrix $\mathbf{I}$, for in that case we have if λ_1, λ_2, ..., are the eigenvalues of $\mathbf{A}^*\mathbf{A}$ and $\boldsymbol{u}_1$, $\boldsymbol{u}_2$, ..., the corresponding eigenvectors:

$$\mathbf{A}^*\mathbf{A}\boldsymbol{u}_k = \lambda_k \boldsymbol{u}_k$$

$$(\mathbf{A}^*\mathbf{A} + \gamma \mathbf{I})\boldsymbol{u}_k = (\lambda_k + \gamma)\boldsymbol{u}_k$$

In other words, the eigenvectors of $A^*A + \gamma I$ are those of A^*A but the eigenvalues are each incremented by γ. Eigenvalues much smaller than γ are greatly increased but those larger than γ are not greatly altered. This in effect filters the oscillatory components, which are associated with the smaller eigenvalues.

Suppose the process of solving a matrix-vector equation $Af = g$ is looked at in terms of the principal axes which the eigenvectors of A provide (we will assume here that A is square). Instead of describing $f(x)$ by the array of value $f(x_1)$, $f(x_2)$, ..., we consider this vector as a combination of the eigenvectors of A, i.e.:

$$f = \sum_k \eta_k u_k = U\eta$$

and to solve the inversion problem in this form we require the elements η_1, η_2, ..., of the vector η, where:

$$g = Af = AU\eta$$

The eigenvalue equation $AU = U\Lambda$ enables the last equation to be written in the form:

$$g = U\Lambda\eta$$

which on direct inversion gives:

$$\eta = \Lambda^{-1}U^{-1}g$$

This formal inversion of course is equivalent to $f = A^{-1}g$ and has no more practical utility than the latter. But it does serve a useful purpose in clarifying the nature of constrained solutions. For if V is the matrix adjoint to U (i.e., $V^*U = I$; V will simply be U itself if A is symmetric), the solution process can be seen to be a projection of g on the "axes", provided by the rows of V followed by a reconstruction of the elements of η by dividing these projections by the respective eigenvalues, a process which gives rise to instability if there are any very small eigenvalues.

Suppose now instead of A^{-1} we were to employ $(A + \gamma I)^{-1}$. The latter, as we have seen, possesses the same eigenvectors as A but the eigenvalues are all increased uniformly by an amount γ. The eigen-equation for $A + \gamma I$ is:

$$(A + \gamma I)\,U = U(\Lambda + \gamma I)$$

and hence:

$$\begin{aligned}(A + \gamma I)^{-1} &= [U(\Lambda + \gamma I)\,U^{-1}]^{-1} \\ &= U(\Lambda + \gamma I)^{-1}\,U^{-1}\end{aligned}$$

If this inverse is applied to the vector g the result is:

$$\begin{aligned}f' = (A + \gamma I)^{-1}\,g &= U(\Lambda + \gamma I)^{-1}\,U^{-1}\,U\Lambda\eta \\ &= U\{(\Lambda + \gamma I)^{-1}\Lambda\}\,\eta\end{aligned}$$

or, in terms of $\boldsymbol{\eta}$:

$$\eta' = \{(\Lambda + \gamma \mathbf{I})^{-1} \Lambda\} \eta$$

The bracketed matrix is a diagonal one with elements $\lambda_1/(\lambda_1 + \gamma)$, $\lambda_2/(\lambda_2 + \gamma)$, etc. When λ_k is much larger than γ, $\lambda_k/(\lambda_k + \gamma)$ differs very little from λ_k but when λ_k is small, $\lambda_k/(\lambda_k + \gamma)$ is small, tending to zero as λ_k tends to zero. In other words if we take η and strike out all terms with $\lambda_k < \gamma$ but retain all terms with $\lambda_k > \gamma$, we obtain η' or something very close to it — the inversion represented by $(\mathbf{A} + \gamma\mathbf{I})^{-1}\boldsymbol{g}$ essentially deletes from η those elements corresponding to small eigenvalues (compared to γ) which are thereby excessively sensitive to small changes or errors in $\boldsymbol{g}$. This is the filtering process which leads to a stability which increases with increasing γ.

6.6. GEOMETRIC NATURE OF CONSTRAINED LINEAR INVERSION

The derivation of the constrained linear inversion formula [6.4] involved two quadratic forms, one (which will be called Q_1) measuring the residual $\mathbf{A}f - \boldsymbol{g}$ (the difference between the measurement data and the data which our solution will produce); the other quadratic form (which will be called Q_2) measured the overall departure from smoothness of the solution vector f. The minimum or minima attained by Q_2 depends on the nature of H, i.e. which measure of smoothness we select. When $\mathbf{H} = \mathbf{I}$, Q_2 is $\Sigma_k f_k^2$, which has a single absolute minimum when $f_1 = f_2 = f_3 = \ldots, = f_N = 0$ — that is, there is one "point" in the N-dimensional space where Σf_k^2 is a minimum and the solution [6.4] will tend to this as γ is made larger and larger. Other forms of Q_2, for example the second difference expression $\Sigma_i(f_{i+1} - 2f_i + f_{i-1})^2$, do *not* give a single unique minimum, but are zero at more than one point in the vector space. As γ is increased the solution [6.4] will tend to one of these points — usually, but not necessarily, that minimum of Q_2 which is closest to the minimum of Q_1.

Fig. 6.7 attempts to illustrate in two dimensions the situation in N-dimensional vector space. We show in the figure *lines* of constant Q_1 (solid) and of constant Q_2 (broken) corresponding to $(N - 1)$-dimensioned *hypersurfaces* of constant Q_1 or Q_2 in the more general situation. Q_1 attains a minimum at "1", Q_2 attains one at "2"; one Q_2-surface, drawn heavier than the others, represents a certain value $Q_2 = k$ and touches a Q_1-surface for $Q_1 = c$. The point S at which they touch represents a solution to either of the problems: (1) minimize Q_1, subject to the constraint $Q_2 = k$; and (2) minimize Q_2, subject to the constraint $Q_1 = c$.

These two problems are each solved by the solution S. Solution formulae such as [6.4] may appear to be asymmetric with respect to γ, but it should be borne in mind that γ is an *undetermined* multiplier and algebraic solu-

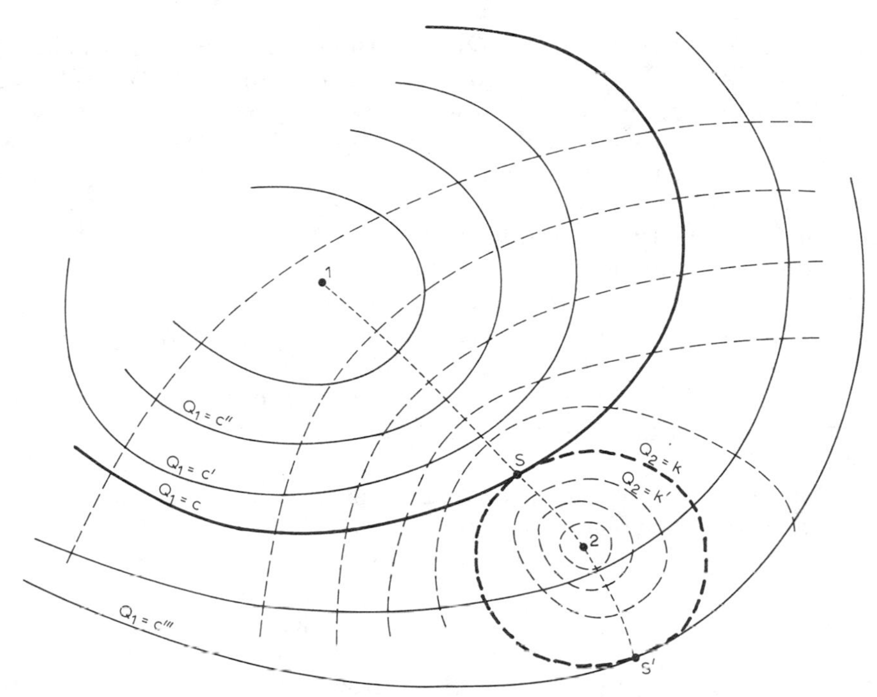

Fig. 6.7. Schematic illustrating solution S and the trajectory in function space (dotted). Continuous curves are level surfaces of Q_1, broken curves level surfaces of Q_2.

tions, for example:

$$f = (\mathbf{A}^*\mathbf{A} + \gamma\mathbf{H})^{-1}\ \mathbf{A}^*g$$

with γ undetermined, are totally equivalent to:

$$f = (\mathbf{H} + \gamma\mathbf{A}^*\mathbf{A})^{-1}\ \mathbf{A}^*g$$

The former gives us a solution to the problem of searching among all possible f's for which $f^*\mathbf{H}f$ = constant for that f which minimizes $|\mathbf{A}f - g|^2$; the latter gives us the result of searching among all possible f's with $|\mathbf{A}f - g|^2$ = constant to find that f for which the smoothness measure $f^*\mathbf{H}f$ is least. The value of the constant is not known — we first locate the solution for some value of the undetermined multiplier γ and only then do we calculate Q_1 or Q_2; this generally must be repeated for several values of γ, obtaining in effect the trajectory of S as γ is varied. As γ goes from zero to infinity S travels along this trajectory from point 1 to point 2 when $Q_1 + \gamma Q_2$ is being minimized. The only effect of using $Q_2 + \gamma Q_1$ is to reverse the direction of movement of S along the trajectory.

The solutions which we have obtained are solutions to the problem: for all f with $|\mathbf{A}f - g|^2 = \epsilon^2$, find the smoothest as judged by $f^*\mathbf{H}f$. If we identify $|\mathbf{A}f - g|^2$ with Q_1, this means that our search for a minimum of Q_2, i.e.,

$f^*\mathbf{H}f$, is restricted to the hypersurface Q_1 = constant (say the heavier solid curve in Fig. 6.7). But if Q_2 should happen to take an equal or *smaller* value somewhere *within* that hypersurface, it would be a better solution, being smoother while giving a lesser residual $Q_1 = |\mathbf{A}f - \mathbf{g}|^2$. What is really required is a solution to the problem: for all f with $Q_1 \leqslant \epsilon^2$, find the smoothest. That is, the search should not be confined to the hypersurface but should extend throughout the interior also. As the heavier curves $Q_1 = c$ and $Q_2 = k$ are drawn in the figure, this minimum evidently *will* lie on the surface and the solution for $Q_1 \leqslant c$ is identical with that for the more restrictive constraint $Q_1 = c$. This is not always true, however; if the surface Q_1 = constant = c''' lies outside of the point 2 (where Q_2 is an absolute minimum) then evidently the solution for $Q_1 \leqslant c'''$ is the point 2, since that point lies in the interior of the region and Q_2 attains its least value there; the solution for $Q_1 = c'''$, however, is the point S', which lies on the surface $Q_1 = c'''$ and on the surface $Q_2 = k$. The solution S' is evidently a false one.

However, it is not bothersome in practice because the behavior of Q_2 with increasing γ is then anomalous and can be recognized immediately. As γ is increased, the value of Q_1 increases while Q_2 first decreases (Q_1 being in the case of equations [6.4], [6.5], etc., the residual or degree of disagreement between $\mathbf{g}$ and $\mathbf{A}f$ or in the case of an integral notation, between $g(y)$ and $\int K(y, x) f(x)\, dx$, while Q_2 is the measure of smoothness).

Once the surface Q_1 = constant is so large that it contains the point 2, Q_1 and Q_2 increase together with increasing γ (see Fig. 6.8). One evidently can only obtain S' as a solution if the solution S (where the surfaces $Q_1 = c$, $Q_2 = k$, touch *externally*) has been missed. The only circumstances that could lead to S' being taken as a "best" solution would be the adoption of excessively coarse steps in γ when a set of values is chosen for that param-

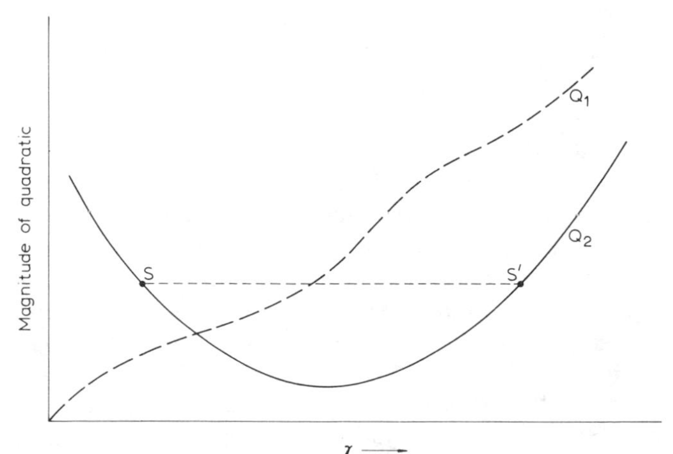

Fig. 6.8. Solution S and false solution S', as related to variation of Q_1 and Q_2.

eter. In most practical situations the trajectory of the solution (dotted line in Fig. 6.7) is confined to the immediate vicinity of the point 1 since usually we demand that $|\mathbf{A}f - \boldsymbol{g}|$ be quite small. If the measurements are so poor or the errors from other sources so great that $|\mathbf{A}f - \boldsymbol{g}|^2$ cannot be confined reasonably tightly — in comparison to the distance in function space between points 1 and 2 — then this means that there is a perfectly smooth solution (e.g. a straight-line relationship) which satisfies the fundamental equation to within the estimated error and we cannot ask for any detailed structural information other than say the mean and slope of $f(x)$ or similar gross properties of that unknown distribution.

BIBLIOGRAPHY

Forsyth, G.E., 1958. Singularity and near singularity in numerical analysis. *Am. Math. Mon.*, 65: 229—240.

Fox, L., 1964. *An Introduction to Numerical Linear Algebra.* Oxford University Press, London, 295 pp.

Gill, P.E. and Murray, F.W., 1974. *Numerical Methods for Constrained Optimization.* Academic Press, New York, N.Y., 283 pp.

Hotelling, H., 1934. Analysis of a complex of statistical variables into principal components. *J. Educ. Psychol.*, 24: 417—441, 498—520.

Householder, A.S., 1964. *The Theory of Matrices in Numerical Analysis.* Blaisdell, New York, N.Y., 257 pp.

Marcus, M., 1960. *Basic Theorems in Matrix Theory (Applied Mathematical Series, N.B.S.).* U.S. Government Printing Office, Washington, D.C., 59 pp.

Murray, F.W., 1972. *Numerical Methods for Unconstrained Optimization.* Academic Press, New York, N.Y., 144 pp.

Ralston, A. and Wilf, H.S., 1960. *Mathematical Methods for Digital Computers.* Wiley, New York, N.Y., 293 pp.

CHAPTER 7

Further inversion techniques

The method of constrained linear inversion was developed to some detail in the last chapter. It was historically one of the first methods to be applied successfully on real inversion problems and is still in widespread use; it is also computationally fast and once a suitable value has been found for the Lagrangian factor γ a whole sequence of observed $\boldsymbol{g}$'s can be processed with only a single matrix-vector multiplication needed for each additional $\boldsymbol{g}$ after the first.

Inversion problems have, however, cropped up to which constrained linear inversion is not well suited. Alternate approaches have been applied sucessfully in some cases and of course there are alternative methods which can be applied in situations where constrained linear inversion techniques are applicable. We will discuss some, but not all, methods in this chapter. It is helpful to categorize a method which gives a solution $f'(x_1)$, $f'(x_2)$, ..., $f'(x_N)$ from an array of measured quantities $g(y_1)$, $g(y_2)$, ..., $g(y_M)$ as linear if the solution is obtained as a linear combination of the measured quantities, i.e. if the vector $\boldsymbol{f}' - [f'(x_1), f'(x_2), ..., f'(x_N)]$ is related to the vector $\boldsymbol{g} = [g(y_1), g(y_2), ..., g(y_M)]$ by a linear transformation:

$$\boldsymbol{f}' = \mathbf{T}\boldsymbol{g} + \boldsymbol{c} \qquad [7.1]$$

i.e.

$$f_i'(x) = \sum_j t_{ij} g(y) + c_i \qquad [7.2]$$

This linear relationship between $\boldsymbol{f}'$ and $\boldsymbol{g}$ may be explicit, as in the case of equation [6.2], [6.3] or [6.4] (see Chapter 6), where in the latter case, for example:

$$\boldsymbol{f}' = (\mathbf{A}^*\mathbf{A} + \gamma\mathbf{H})^{-1}\mathbf{A}^*\boldsymbol{g}$$

is a linear relationship of the form $\boldsymbol{f}' = \mathbf{T}\boldsymbol{g}$, with $\mathbf{T}$ given by $(\mathbf{A}^*\mathbf{A} + \gamma\mathbf{H})^{-1}$ $\mathbf{A}^*\boldsymbol{g}$. In computing a solution by this method, $\mathbf{T}$ is normally not calculated explicitly but the relationship between $\boldsymbol{f}'$ and $\boldsymbol{g}$ is evidently linear. Methods which are less direct, including iterative methods may be linear and often are, even though a cursory inspection may not suggest it.

Non-linear methods are also employed and in some situations appear to be preferable to linear methods. A non-linear method is typified by a relationship:

$$f' = \Phi g \tag{7.3}$$

where Φ depends on $\boldsymbol{g}$ or on some of its elements or some linear combination thereof. To apply non-linear methods to obtain solutions for a sequence of $\boldsymbol{g}$'s one must start afresh from the beginning each time since a different Φ will be involved. Generally they are considerably slower than linear methods, so there is a price to be paid for the adoption of non-linear methods.

7.1. MORE ELABORATE TREATMENTS OF ERROR COMPONENTS IN LINEAR INVERSIONS

Errors as we have seen are crucial in inversion problems and to obtain the best possible solution without sacrificing stability it is necessary to categorize the errors in the measurements $g_1, g_2, ..., g_M$ as accurately as possible. By their very nature errors are statistical in character and there is always an element of uncertainty — it may be concluded that the magnitude of an error $|\epsilon|$ is unlikely ever to exceed some prescribed value, but there will always be some small but non-zero chance that it will be larger than that value, and an inversion based on that limit will occasionally fail for the same reason. However, it is important that the known properties of the errors be assessed as far as is possible and incorporated into the inversion procedure.

The concept of an error such as the error ϵ in a measurement $\boldsymbol{g}$ implies that the measurement gives a neighborhood within which the true $\boldsymbol{g}$ is likely to be found; the neighborhood does not have definite mathematical bounds — indeed it extends from $-\infty$ to $+\infty$ in most situations — but as we go away from the measured value g_k the probability of the true g_k being found at a given point diminishes. If we knew the true value of g_k and made a sequence of measurements of this quantity, g_k', g_k'', g_k''', etc., we would be able to construct a table or graph showing the distribution of the error; in most cases we do not know this distribution and can only estimate the average magnitude of the error, but the normal or Gaussian distribution is often found in measured quantities and is usually assumed when the distribution is symmetric, i.e. when positive or negative errors are equally likely. When the distribution is symmetric the mean error is zero and the root mean square (r.m.s.) error is used to describe the magnitude of the error.

Errors may be envisaged as being sampled in a random way by drawing from a "bag" of errors; we have an idea (often quite a good one) of the r.m.s. error but do not of course know a priori the sign, much less the value, which will be drawn from the bag in a given draw. When there are several measurements we have of course several errors and an error vector which is drawn from a "bag" of error vectors; it is important to note that this is not necessarily the same as drawing at random the individual components of the error vector from a bag of random scalar errors. Consider the array:

$$
\begin{array}{cccc}
\epsilon_1' & \epsilon_1'' & \epsilon_1''' & \dots \\
\epsilon_2' & \epsilon_2'' & \epsilon_2''' & \dots \\
\epsilon_3' & \epsilon_3'' & \epsilon_3''' & \dots \\
. & . & . & \\
. & . & . & \\
\epsilon_m' & \epsilon_m'' & \epsilon_m''' & \dots
\end{array}
$$

which specifies completely a set of samplings of m-dimensioned error vectors. The mean of the squares of the elements of the first row is the r.m.s. error of the first measurement, which is usually known or can at least be estimated; this quantity is also known for the other rows. Consider now the $m \times m$ matrix **E** obtained by forming the inner (dot) product of the various pairs of rows and dividing by the number of samplings, that is:

$$\mathbf{E} = s^{-1} \| \epsilon_i' \epsilon_j' + \epsilon_i'' \epsilon_j'' + \dots + \epsilon_i^{''\cdots'} \epsilon_j^{''\cdots'} \| \qquad [7.4]$$

When s is large enough **E** tends to a definite limit; if the i- and j-errors are independent of each other, the signs in the sum $\epsilon_i' \epsilon_j' + \epsilon_i'' \epsilon_j'' + \dots$, will alternate in a random way and the corresponding elements will be zero. The diagonal elements are of course $s^{-1}[(\epsilon_i')^2 + (\epsilon_i'')^2 + \dots,]$, which is the mean square value for the error ϵ_i, which will not vanish; furthermore if we scale the problem so that the mean square errors are all equal (such a scaling is always possible and is indeed desirable in practice), then Schwarz's inequality ensures that the off-diagonal elements cannot exceed the diagonal elements, since:

$$\epsilon_i' \epsilon_j' + \epsilon_i'' \epsilon_j'' + \dots \leqslant [(\epsilon_i')^2 + (\epsilon_i'')^2 + \dots]^{1/2} [(\epsilon_j')^2 + (\epsilon_j'')^2 + \dots]^{1/2}$$

Equality results only when $\epsilon_i' = \epsilon_j'$, $\epsilon_i'' = \epsilon_j''$, etc., so that ϵ_i and ϵ_j are perfectly correlated with each other.

The matrix **E** therefore prescribes not just the mean square errors but also the extent to which these errors are interdependent (correlated statistically). If the errors are independent **E** is a diagonal matrix; if their mean square values are all equal then the diagonal elements of **E** are all equal. With uncorrelated errors with equal r.m.s. values e, **E** is simply $e^2\mathbf{I}$.

If we know **E** and it contains some non-zero off-diagonal elements, then this implies that it is not correct to regard all error vectors with the same norm $|\epsilon|^2$ as equally probable. The existence of non-zero off-diagonal elements, say in the 1, k position, implies that ϵ_1 and ϵ_k are not independent of each other. If E_{1k} is positive ϵ_1 tends to be positive when ϵ_k is positive

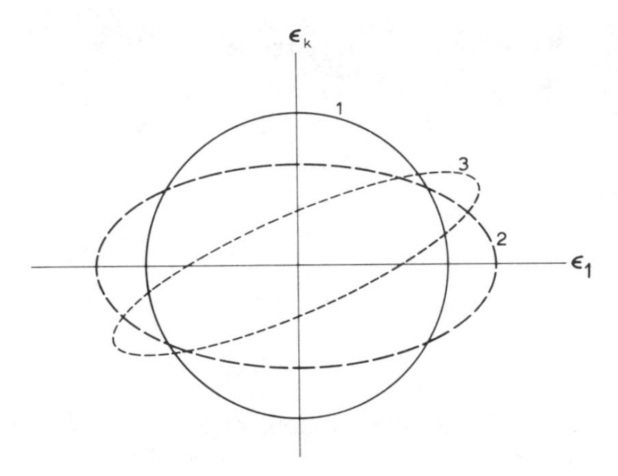

Fig. 7.1. Curves of constant probability for: (1) equal and independent, (2) unequal and independent, and (3) unequal and correlated errors.

and negative when ϵ_k is negative, while if E_{1k} is negative ϵ_1 and ϵ_k prefer to take opposite signs.

If we look only at ϵ_1 and ϵ_k, a diagram can be plotted showing curves of constant probability in the (ϵ_1, ϵ_k) plane. When ϵ_1 and ϵ_k are independent and have equal r.m.s. values the equal-probability curve is a circle (Fig. 7.1, solid curve); if ϵ_1 has a larger expectation (larger r.m.s. value) than ϵ_k the curve becomes an ellipse, with axes coinciding with the ϵ_1 and ϵ_k axes (Fig. 7.1, broken curve); but if there is a positive correlation between ϵ_1 and ϵ_k the curve becomes an ellipse with its major axis tilted (dotted curve) — the quadrant-to-quadrant symmetry which previously existed has been destroyed by the correlation which makes the quadrants $(\epsilon_1 > 0, \epsilon_k > 0)$ and $(\epsilon_1 < 0, \epsilon_k < 0)$ preferred against the other two quadrants. If ϵ_1 and ϵ_k are totally correlated quadrants we have $\epsilon_1 \equiv \alpha\epsilon_k$ and no curve exists; the ellipse degenerates into a straight line.

Modifications to the solution process in the presence of unequal or partially interdependent errors

In the derivation of the inversion equation [6.4], it was prescribed that the errors $\epsilon_1, \epsilon_2, ..., \epsilon_M$ were related by the requirement:

$$\sum_k \epsilon_k^2 = e^2$$

but otherwise were independent — in other words the errors were given equal weight and assumed to be random, independent quantities with the same expectation. The only constraint was that admissible error vectors are confined to a sphere in N-dimensional vector space.

Unequal error expectations. In some instances some error components may be systematically larger than others. Very little extra algebra is needed to allow for such considerations — one can either rescale the components of f or obtain a form of [6.4] in which, for example, the matrix $\mathbf{I}$ is replaced by a matrix which is still diagonal but with diagonal elements which are unequal.

The first form emerges as follows. Suppose weights $w_1, w_2, \ldots, w_M$ are assigned which are proportional to the expected magnitude of the various errors, then the quantities ϵ_1/w_1, ϵ_2/w_2, etc., will now represent samplings from the same population and possess equal variances. Hence, if we rewrite:

$$\mathbf{A}f = g + e$$

as:

$$\mathbf{W}^{-1}\mathbf{A}f = \mathbf{W}^{-1}g + \mathbf{W}^{-1}e \qquad (W_{kk} = w_k;\ W_{kj} = 0,\ k \neq j) \tag{7.5}$$

we have restored the equation to a form in which the expected magnitude of the N error components is again the same and we can legitimately solve that equation by using [6.4], putting $\mathbf{W}^{-1}\mathbf{A}$ in place of $\mathbf{A}$ and $\mathbf{W}^{-1}g$ in place of g; this gives:

$$f = [(\mathbf{W}^{-1}\mathbf{A})^*\mathbf{W}^{-1}\mathbf{A} + \gamma\mathbf{H}]^{-1}\ (\mathbf{W}^{-1}\mathbf{A})^*\mathbf{W}^{-1}g$$

or:

$$f = [\mathbf{A}^*\mathbf{W}^{-1}\mathbf{W}^{-1}\mathbf{A} + \gamma\mathbf{H}]^{-1}\ \mathbf{A}^*\mathbf{W}^{-1}\mathbf{W}^{-1}g \tag{7.6}$$

$\mathbf{W}$ of course is a diagonal matrix with elements $w_1, w_2, \ldots, w_M$.

The second method is to use explicitly the constraint $\sum_i \epsilon_i^2\, w_i^{-2}$ = constant, so that we minimize $f^*\mathbf{H}f$ subject to the constraint $(\mathbf{W}^{-1}\epsilon)^*\ (\mathbf{W}^{-1}\epsilon)$ = constant, that is, we find f so that:

$$\frac{\partial}{\partial f}(f^*\mathbf{H}f) = 0 \qquad (k = 1, 2, \ldots N)$$

while $[\mathbf{W}^{-1}(\mathbf{A}f - g)]^*[\mathbf{W}^{-1}(\mathbf{A}f - g)]$ = constant.

The last expression is still a quadratic form Q in f so the nature and form of our solution will obviously be unchanged. The quadratic form can be expanded:

$$Q = f^*\mathbf{A}^*\mathbf{W}^{-1}\mathbf{W}^{-1}\mathbf{A}f - g^*\mathbf{W}^{-1}\mathbf{W}^{-1}\mathbf{A}f - f^*\mathbf{A}^*\mathbf{W}^{-1}\mathbf{W}^{-1}g + g^*\mathbf{W}^{-1}\mathbf{W}^{-1}g$$

The last term is independent of f and can be dropped. The vanishing of the derivatives of the sum of Q and $\gamma f^*\mathbf{H}f$ implies that:

$$e^*\mathbf{A}^*\mathbf{W}^{-1}\mathbf{W}^{-1}\mathbf{A}f + f^*\mathbf{A}^*\mathbf{W}^{-1}\mathbf{W}^{-1}\mathbf{A}e - g^*\mathbf{W}^{-1}\mathbf{W}^{-1}\mathbf{A}e - e^*\mathbf{A}^*\mathbf{W}^{-1}\mathbf{W}^{-1}g$$
$$+ \gamma e^*\mathbf{H}f + \gamma f^*\mathbf{H}e = 0 \qquad (k = 1, 2, \ldots N)$$

which reduces to:

$$(\mathbf{A}^*\mathbf{W}^{-1}\mathbf{W}^{-1}\mathbf{A} + \gamma\mathbf{H})f = \mathbf{A}^*\mathbf{W}^{-1}\mathbf{W}^{-1}g$$

or:

$$f = (\mathbf{A}^*\mathbf{W}^{-1}\mathbf{W}^{-1}\mathbf{A} + \gamma\mathbf{H})^{-1}\mathbf{A}^*\mathbf{W}^{-1}\mathbf{W}^{-1}g$$

which is just equation [7.6]. Hence, as we would expect, both methods give the same final solution. The formula contains expressions such as $\mathbf{A}^*\mathbf{W}^{-1}$-$\mathbf{W}^{-1}\mathbf{A}$, which in component form are:

$$\sum_k a_{ki} w_k^{-2} a_{kj}$$

which could be regarded as resulting from a redefinition of the inner product.

Correlated errors. The assumption of independent errors implies that there is no correlation between any pair of errors, so that in summing over a large number of observations a pair of values such as $\epsilon_i = 0.7$, $\epsilon_j = -0.5$ will be canceled by a pair of values $\epsilon_i = 0.7$, $\epsilon_j = 0.5$ or $\epsilon_i = -0.7$, $\epsilon_j = -0.5$. We also have assumed here, as previously, that all errors have mean zero — any known systematic component can be removed and we do not know the sign of an unknown systematic component so this is not at all a restrictive assumption.

When there are known to be correlations among the errors it is obviously desirable to take these correlations into account in a solution. If ϵ_i and ϵ_j "want" to have the same sign then clearly the solution should be biased against combinations containing opposite signs in those positions — they are less likely but they can occur and cannot just be eliminated.

What is required to do this is simply replacement of the criterion:

$$\epsilon^*\epsilon = \sum_i \epsilon_i^2 = \text{constant}$$

by a similar but more complex criterion which takes into account the statistical ties between the error elements. That is, when we search through the vector space of all possible ϵ to find a solution which minimizes a measure of smoothness in the functional $f^*\mathbf{H}f$ (which varies as ϵ is varied), we do not give the same consideration to all ϵ's which have the same length, but give preference to those which are statistically favored. The surfaces of equal a priori probability in the ϵ-vector space are then no longer spheres but can be transformed back into spheres if we find a suitable transformation:

$$\zeta = \mathbf{L}\epsilon$$

that is, the constraint $|\epsilon|^2 = \epsilon^*\epsilon = \text{constant}$ must be replaced by a constraint $|\zeta|^2 = \zeta^*\zeta = \text{constant}$. When we sample any ϵ there is paired with it a vector ζ in ζ-space and vice versa. The surfaces of equal probability in ζ-space are to be hyperspheres so that if we randomly sample ζ's we find just as many positive combinations as negative and over a large enough sample $\overline{\zeta_k\zeta_j} = 0$ when k

$\neq j$, while $\overline{\zeta_k^2} = c^2$ for all values of k. But the elements $\zeta_k \zeta_j$ and ζ_k^2 are contained in the matrix:

$$\zeta\zeta^* = \mathbf{L}\epsilon\epsilon^*\mathbf{L}^*$$

Averaged over many samples the off-diagonal elements of the matrix $\zeta\zeta^*$ are to vanish and the diagonal elements to become c^2, i.e. we require that $\zeta = \mathbf{L}\epsilon$ be such that:

$$\overline{\zeta\zeta^*} = c^2\mathbf{I}$$

but:

$$\overline{\zeta\zeta^*} = \mathbf{L}\overline{\epsilon\epsilon^*}\,\mathbf{L}^* = \mathbf{LEL}^*$$

$\mathbf{E}$ being the error covariance matrix mentioned earlier; hence one requires:

$$\mathbf{LEL}^* = c^2\mathbf{I}$$

But if $\mathbf{U}$ is the matrix of eigenvectors of $\mathbf{E}$ and $\Lambda^{1/2}$ the diagonal matrix with elements $\lambda_1^{1/2}$, $\lambda_2^{1/2}$, etc., then the familiar relationship:

$$\mathbf{U}^*\mathbf{EU} = \Lambda$$

gives us the procedure for "diagonalizing" the error vectors. If we choose for $\mathbf{L}$ the matrix $\Lambda^{-1/2}\mathbf{U}^*$ (which incidentally is not symmetric), the vector $\zeta = \mathbf{L}\epsilon$ displays the required statistical properties, for then we have:

$$\begin{aligned}\overline{\zeta\zeta^*} &= \overline{\Lambda^{-1/2}\,\mathbf{U}^*\epsilon\epsilon^*\mathbf{U}\,\Lambda^{-1/2}} \\ &= \Lambda^{-1/2}\mathbf{U}^*\mathbf{EU}\,\Lambda^{-1/2} = \Lambda^{-1/2}\,\Lambda\,\Lambda^{-1/2} = \mathbf{I}\end{aligned}$$

If we know $\mathbf{E}$ (which in most practical situations is easier said than done) then the most appropriate constrained solution is obtained by applying the constraint $|\zeta|^2$ = constant rather than $|\epsilon|^2$ = constant, so that instead of searching over the surface of a hypersphere in ϵ-space we search over a hypersphere in ζ-space (which is equivalent to searching over an ellipsoid in ϵ-space).

The problem of minimizing $f^*\mathbf{H}f$ subject to the constraint $\zeta^*\zeta$ = constant is no more complicated than the problems treated earlier: $\zeta^*\zeta$ is still a quadratic form in f since $\zeta = \mathbf{L}\epsilon = \mathbf{L}(\mathbf{A}f - g)$, or:

$$\begin{aligned}\zeta^*\zeta &= (f^*\mathbf{A}^* - g^*)\mathbf{L}^*\mathbf{L}(\mathbf{A}f - g) \\ &= f^*\mathbf{A}^*\mathbf{L}^*\mathbf{LA}f - g^*\mathbf{L}^*\mathbf{LA}f - f^*\mathbf{A}^*\mathbf{L}^*\mathbf{L}g + g^*\mathbf{L}^*\mathbf{L}g\end{aligned}$$

The constrained solution is given as usual by:

$$\frac{\partial}{\partial f_k}\{f^*\mathbf{A}^*\mathbf{L}^*\mathbf{LA}f - f^*\mathbf{A}^*\mathbf{L}^*\mathbf{L}g - g^*\mathbf{L}^*\mathbf{LA}f + \gamma f^*\mathbf{H}f\} = 0 \qquad (k = 1, 2, \ldots N)$$

which differs from the equations found earlier only to the extent that $\mathbf{A}$ has been replaced by **LA**, i.e. $\Lambda^{-1/2}U^*A$, and $\boldsymbol{g}$ by $\mathbf{L}\boldsymbol{g}$. The solution is given by [6.5] if we insert in place of **A** the above matrix, i.e. it is:

$$(A^*U\Lambda^{-1}U^*A + \gamma H)f = A^*U\Lambda^{-1/2}\Lambda^{-1/2}U^*g$$

or:

$$f = (A^*L^*LA + \gamma H)^{-1} A^*L^*Lg \qquad [7.7]$$

a straightforward modification of equation [6.5]. Since γ is undetermined there is no loss of generality if for **L** we use the simplest expression $\Lambda^{-1/2}U^*$; L^*L is then $U\Lambda^{-1}U^* = E^{-1}$ and so [7.7] can be written:

$$f = (A^*E^{-1}A + \gamma H)^{-1} A^*E^{-1}g \qquad [7.8]$$

a form which can be regarded formally as a direct application of equation [6.5] when the fundamental equation $\mathbf{A}f = g + \epsilon$ is first multiplied by **L**, giving:

$$\mathbf{LA}f = \mathbf{L}g + \mathbf{L}\epsilon = \mathbf{L}g + \epsilon'$$

the transformation **L** being chosen so as to make the elements of the transformed error vector ϵ' mutually independent and possessing the same expected magnitude. As was mentioned earlier it is not often that the statistical behavior of the errors including their mutual interdependences are completely known, to the extent that **E** can be prescribed. Once we assume independence of the various error components, equation [7.8] reverts to the earlier solution [6.5] for then **E** becomes the identity matrix.

7.2. THE SYNTHESIS APPROACH TO INVERSION

All linear methods of solution give (implicitly or explicitly) a solution $f'(x_i)$, $i = 1, 2, ..., n$, which, as was mentioned above, can be written:

$$f'(x_i) = \sum_j t_{ij} g_j$$

Since g_j is $\int K_j(x)\, f(x)\, \mathrm{d}x$, plus an error term which will be dropped for the moment, the solution can be written:

$$f'(x_i) = \sum_j t_{ij} \int K_j(x)\, f(x)\, \mathrm{d}x$$

$$= \int \sum_j t_{ij} K_j(x) \cdot f(x)\, \mathrm{d}x$$

So:

$$f'(x_i) = \int s_i(x)\, f(x)\, \mathrm{d}x$$

where the combination $\Sigma_j \; t_{ij} \; K_j(x)$ has been defined to be a "scanning function" $s_i(x)$. The only way in which $f'(x_i)$ can exactly equal $f(x_i)$ is for the scanning function to be the delta function $\delta(x - x_i)$ centered at $x = x_i$. In that event:

$$f'(x_i) = \int \delta(x - x_i) \, f(x) \, dx = f(x_i)$$

Thus the problem of inverting the equation $\boldsymbol{g} = \int \boldsymbol{k}(x) \, f(x) \, dx$ can be viewed as one of synthesizing a delta function by a linear combination of kernels. There is, of course, no finite set of coefficients which will give a delta function precisely, but we can gain a pretty good picture of how $f(x)$ behaves if we can look at it through a "window" which is not too wide — even if it is not the idealized slit of infinitesimal width represented by the delta function. Indeed every realizable *measurement* "at" $x = x_i$ gives one not $f(x_i)$ but $\int s(x - x_i) \, f(x) \, dx$ where $s(x - x_i)$ — a "slit function" in spectroscopic parlance — is largest near x_i, but does not vanish identically when $x \neq x_i$. A solution to an inversion problem would be perfectly acceptable if it gave us a similar "smearing" of $f(x)$, and it is therefore relevant to look for a linear combination of kernel functions which, while not a delta function, is nevertheless a good approximation to a delta function. In other words, in place of the self-evident relationships:

$$f(x_1) = \int \delta(x - x_1) \, f(x) \, dx$$

$$f(x_2) = \int \delta(x - x_2) \, f(x) \, dx$$

$$\cdot$$

$$\cdot$$

$$f(x_N) = \int \delta(x - x_N) \, f(x) \, dx$$

one seeks combinations of kernels to approximate each delta function:

$$\delta(x - x_1) \cong t_{11}K_1(x) + t_{12}K_2(x) + \ldots + t_{1M}K_M(x)$$

$$\delta(x - x_2) \cong t_{21}K_1(x) + t_{22}K_2(x) + \ldots + t_{2M}K_M(x)$$

$$\cdot$$

$$\cdot$$

$$\delta(x - x_N) \cong t_{N1}K_1(x) + t_{N2}K_2(x) + \ldots + t_{NM}K_M(x)$$

and thereby arrives at approximate solutions (again indicated by primes):

$$f'(x_1) = \sum_k t_{1k} g_k \cong f(x_1)$$

$$f'(x_2) = \sum_k t_{2k} g_k \cong f(x_2)$$

.

.

$$f'(x_N) = \sum_k t_{Nk} g_k \cong f(x_N)$$

This inversion is essentially linear and fundamentally differs little from other linear solutions. It is intuitively apparent that if the quadrature formula $g \cong \mathbf{A}f$ gives very large elements in $\mathbf{A}^{-1}$ when inverted to obtain the direct inverse $f = \mathbf{A}^{-1}g$, then the elements of $\mathbf{T}$ are also likely to be very large, and this is indeed found in practice. One therefore finds that when g contains small error components, the solution for any $f(x_i)$, being given by:

$$f'(x_i) = \sum_k t_{ik} g_k + \sum_k t_{ik} \epsilon_k$$

is as usual dominated by the error term and very sensitive to very small changes in any of the g_k, since $(\partial/\partial g_k)\, f'(x_i) = t_{ik}$, which may be very large and of either sign.

Up to this point the synthesis approach has given a new perspective to the inversion problem but it has not succeeded in eliminating the fundamental difficulties — we are still confronted with instability and error magnification.

Quite obviously to obtain a usable inversion formula the elements of T must be limited so as to keep the magnitude of the error terms $\Sigma_k\ t_{ik}\ \epsilon_k$ within reasonable bounds. This can be done in a very similar way to that used in Chapter 6 to limit the magnitude of f or of some linear combination of its elements. This is the basis of an inversion technique known as the Gilbert-Backus method.

The Gilbert-Backus method

The Gilbert-Backus method applies a quadratic constraint to the elements of $\mathbf{T}$ so as to control the error contribution $\Sigma_k\ t_{ik}\ e_k$; furthermore, the scanning functions $s_1(x)$, $s_2(x)$, ..., are each prescribed to possess the normalization property $\int s_i(x)\ dx = 1$ — this is clearly necessary if a constant function is to be unaltered by the smoothing process represented by the operation $\int s_i(x)\, f(x)\ dx$, and is simply a necessary scaling requirement.

The solution obtains each row of $\mathbf{T}$ separately, with the following constraints (writing $\boldsymbol{r}$ to denote the row of $\mathbf{T}$ which is being calculated, deleting a subscript i to keep the algebraic notation simple):

(1) $\boldsymbol{r}^*\boldsymbol{r} \leqslant$ constant, to prevent inordinate error magnification.

(2) $\int s_i(x)\ dx = \boldsymbol{r}^* \int \boldsymbol{k}(x)\ dx = \int \boldsymbol{k}^*(x)\ dx\ \boldsymbol{r} = 1$, to make the scanning function unitary.

(3) a requirement that $s_i(x)$ be "concentrated" in some fashion around $x = x_i$ so that it approximates in some sense a delta function centered at $x = x_i$. This latter requirement gives a degree of flexibility since different measures can be found to gauge the degree to which the function $s_i(x)$ picks out the immediate neighborhood of x_i. One possible choice is the square norm $\int [s_i(x) - \phi_i(x)]^2 \, dx$ where $\phi_i(x)$ is a function approximating or tending to the delta function $\delta(x - x_i)$. Another possible choice and one which is favored by the originators of the method, is the "spread" $\int (x - x_i)^2 [s_i(x)]^2 \, dx$; other choices are obviously possible — for example, one might choose a higher power than the second to measure the "spread" of $s_i(x)$ — but there is not much point in using them unless the special nature of a specific problem calls for special treatment.

The measures just given are both quadratic forms in the vector $\boldsymbol{r}$ which constitutes the solution to the problem for the ith row of $\mathbf{T}$. If the solution $f'(x)$ is to be obtained at a sequence of values of x_i the complete solution process must be repeated for each of these. The reader anticipating a matrix inversion to obtain $\boldsymbol{r}$ is correct, and it is worth noting that a separate matrix inversion will be required for each i, so a solution $f'(x_i)$, $i = 1, 2, \ldots, n$, to the inversion problem requires that the solution process for $\boldsymbol{r}$ (which we are about the describe) be repeated n times, each row or set of coefficients for $i = 1, 2, \ldots,$ being obtained quite independently of the others.

To obtain a single row $\boldsymbol{r}$ we evidently must solve a slight modification of a very familiar problem — what is required here is the minimization of a quadratic form $\boldsymbol{r}^* \mathbf{S} \boldsymbol{r}$ in $\boldsymbol{r}$, subject to a pair of constraints:

$$\boldsymbol{r}^* \boldsymbol{r} = \text{constant}$$

and

$$\boldsymbol{b}^* \boldsymbol{r} = \boldsymbol{r}^* \boldsymbol{b} = 1$$

$\boldsymbol{b}$ is written to represent the vector $\int \boldsymbol{k}(x) \, dx$, i.e. $b_i = \int K_i(x) \, dx$. The latter constraint is simply the normalization constraint $\int s_i(x) \, dx = 1$.

The extremum condition is given by:

$$\frac{\partial}{\partial r_k} [\boldsymbol{r}^* \mathbf{S} \boldsymbol{r} - \beta \boldsymbol{b}^* \boldsymbol{r} - \beta \boldsymbol{r}^* \boldsymbol{b} + \gamma \boldsymbol{r}^* \boldsymbol{r}] = 0$$

Here there are two Lagrangian multipliers $-\beta$, γ. The two terms $\beta \boldsymbol{b}^* \boldsymbol{r} + \beta \boldsymbol{r}^* \boldsymbol{b}$ are written in this way for symmetry; this sum equals either $2\beta \boldsymbol{b}^* \boldsymbol{r}$ or $2\beta \boldsymbol{r}^* \boldsymbol{b}$ so strictly the Lagrangian multiplier is -2β rather than $-\beta$. The reason for using -2β will immediately be apparent.

Proceeding in the usual way, one finds:

$$\boldsymbol{e}^* \mathbf{S} \boldsymbol{r} + \boldsymbol{r}^* \mathbf{S} \boldsymbol{e} - \beta \boldsymbol{b}^* \boldsymbol{e} - \beta \boldsymbol{e}^* \boldsymbol{b} + \gamma \boldsymbol{e}^* \boldsymbol{r} + \gamma \boldsymbol{r}^* \boldsymbol{e} = 0$$

and hence:

$$\mathbf{S} \boldsymbol{r} + \gamma \boldsymbol{r} = \beta \boldsymbol{b}$$

$$\left.\begin{aligned} \boldsymbol{r} &= \beta(\mathrm{S} + \gamma\mathrm{I})^{-1}\boldsymbol{b} \\ s_i(x) &= \beta\boldsymbol{k}^*(x)\,(\mathrm{S} + \gamma\mathrm{I})^{-1}\boldsymbol{b} \end{aligned}\right\} \qquad [7.9a]$$

The magnitude of β is readily determined since we have a constraint $\boldsymbol{r}^*\boldsymbol{b} = \boldsymbol{b}^*\boldsymbol{r} = 1$. This gives:

$$\beta\boldsymbol{b}^*(\mathrm{S} + \gamma\mathrm{I})^{-1}\boldsymbol{b} = 1$$

so that β is given by:

$$\beta = \frac{1}{\boldsymbol{b}^*(\mathrm{S} + \gamma\mathrm{I})^{-1}\boldsymbol{b}} \qquad [7.10]$$

the denominator of the fraction is, of course, a scalar quantity — a quadratic form in the known vector $\boldsymbol{b} = \int \boldsymbol{k}(x)\,\mathrm{d}x$.

Equations [7.9a] and [7.10] form the basis of the Gilbert-Backus procedure. Note that $\boldsymbol{b}$, being independent of i, is the same for each row, but S, being the matrix of the quadratic form which measures the departure of $s_i(x)$ from the delta function $\delta(x - x_i)$ *does* depend on i, so the matrix inversion $(\mathrm{S} + \gamma\mathrm{I})^{-1}$ must be recomputed for each value of i.

The final solution can of course be written explicitly since it is the result of operating on $f(x)$ by the scanning function $s_i(x)$, i.e. the solution at $x = x_i$ is:

$$\begin{aligned} f'(x_i) &= \int s_i(x)\,f(x)\,\mathrm{d}x \\ &= \int \boldsymbol{r}^*\boldsymbol{k}(x)\,f(x)\mathrm{d}x = \boldsymbol{r}^*\boldsymbol{g} \end{aligned} \qquad [7.9b]$$

Each $\boldsymbol{r}^*$ can be put into a row position of a matrix and the solution for several tabular x_i written as one matrix equation, but there is no particular value to such a form in this instance, since the rows of the matrix are given by separate matrix-vector equations.

The selection of γ in the Gilbert-Backus procedure is similar to that in the constrained inversion methods of the previous chapter. As γ is decreased the scanning functions become narrower, but the price paid is the increase in the elements of $\boldsymbol{r}$ (i.e. t_{ij}, $j = 1, 2, ..., n$) which become larger with decreasing γ. The error magnification depends on the magnitude of these elements:

error in $\boldsymbol{g} = \epsilon$

error component in $f'(x) = \boldsymbol{r}^*\epsilon$

error magnification $= |\boldsymbol{r}^*\epsilon|/|\epsilon| = \sqrt{\dfrac{r_1^2\epsilon_1^2 + r_2^2\epsilon_2^2 + ...}{\epsilon_1^2 + \epsilon_2^2 + ...}}$

The worst error magnification is the largest element of $\boldsymbol{r}$ and an average r.m.s. magnification for uncorrelated errors is $m^{-1/2}|\boldsymbol{r}|$. We can therefore for each γ obtain a value for the spread of the scanning function and an error

TABLE 7.1

Error magnification and width of scanning function

γ	Error magnification	Width of scanning function
10^{-2}	2.5	0.85
10^{-4}	5.8	0.62
10^{-6}	52	0.45
10^{-8}	370	0.41
10^{-10}	2800	0.33
10^{-12}	12000	0.31

magnification; these can be plotted against each other as a trade-off curve which essentially shows the price which must be paid in terms of accuracy to obtain a given spread.

For $x_i = 0.4$ and the $x\ e^{-yx}$ kernels which have been used in earlier problems, the data shown in Table 7.1 were obtained. Beyond the γ values in the last line (see Table 7.1) very little improvement in the width of the scanning function was obtained, but the error magnification continued to grow. The width as given above was the "half-width" of the scanning function, i.e. the distance between the points where the amplitude had fallen to one-half of its value at the peak of the curve. This is not the same measure of "spread" as that used to derive the mathematical solution but it furnishes a numerical description of the spread which is simple to grasp. (The half-width of $s_i(x)$ cannot be expressed as a quadratic form in the array of coefficients $\boldsymbol{r}$ and so it cannot be minimized by means of equation [7.9a]).

The type of data tabulated above are a direct result of applying equation [7.9a] and they give a very useful and readily comprehended description of the effectiveness and resolving power of particular indirect sensing measurements. Unfortunately the limitations which beset the inversion problem are still present — as one can see from Table 7.1 and from Figs. 7.2 and 7.3 where synthesized scanning functions and trade-off curves are shown for a few representative values of x. We will not go further into the question of resolving power here as this important subject will be dealt with specifically in a later chapter, but it is evident that the resolving power of the inversion is quite restricted for any γ.

Quadratic measures of the spread of $s_i(x)$

Equation [7.9a] can be applied to synthesize a scanning function $s_i(x)$ so as to minimize any given positive definite quadratic form $\boldsymbol{r}^*\mathrm{S}\boldsymbol{r}$. We have only to specify the form of S, and that is a simple task for simply defined measures of spread. In particular the "spread" $\int(x - x_i)^2\ s_i^2(x)\ dx$, which was

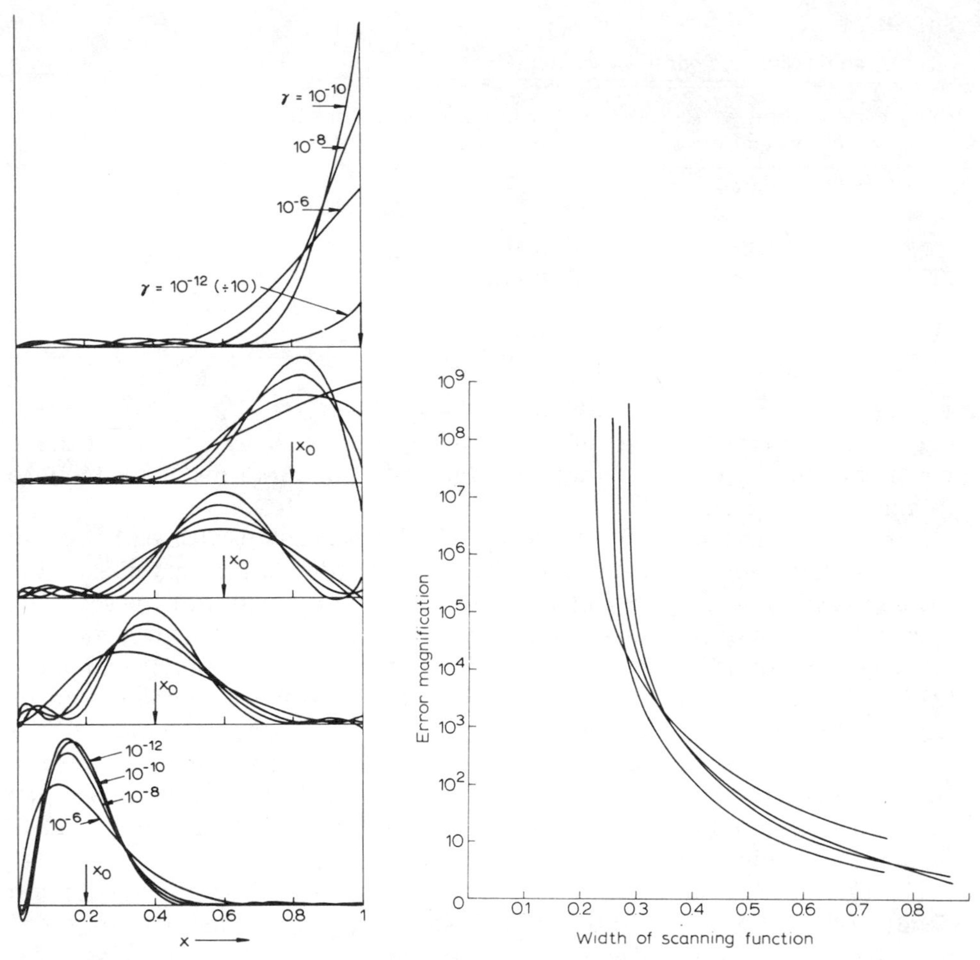

Fig. 7.2. Scanning functions synthesized from the kernel functions $x\ e^{-yx}$ by means of equation [7.9a]; $y = 20, 10, 6.67, 5, 4, ..., 1$. Arrows indicate x_0, the central value (about which spread of the scanning functions was minimized).

Fig. 7.3. "Trade-off" curves for the Gilbert-Backus procedure, applied to the same kernels as were used in Fig. 7.2. Different curves relate to different abscissa values for the center of the scanning function.

the one used to derive the results already given can be expressed as a quadratic form as follows:

We have:

$$s_i(x) = \boldsymbol{r}^* \boldsymbol{k}(x) = \boldsymbol{k}^*(x)\boldsymbol{r} \qquad \text{so: } [s_i(x)]^2 = \boldsymbol{r}^* \boldsymbol{k}(x)\, \boldsymbol{k}^*(x)\boldsymbol{r}$$

and:

$$\int (x - x_i)^2\, s_i^2(x)\, \mathrm{d}x = \boldsymbol{r}^* \{ \int (x - x_i)^2 \boldsymbol{k}(x) \boldsymbol{k}^*(x) \mathrm{d}x \} \boldsymbol{r}$$

The bracketed term is the matrix S required. Inserted into equation [7.9a] this particular S leads to a scanning function $s_i(x)$ which has unit area ($\int s_i(x)\,dx = 1$), is a linear combination of the kernels $K_1(x)$, $K_2(x)$, etc., and of all such combinations gives the least value of $\int (x - x)^2\, s_i^2(x)\,dx$, subject to the constraint which limits $\boldsymbol{r}^*\boldsymbol{r}$ (i.e. the overall magnitude of the coefficients), and thereby the error magnification.

Another quadratic spread measure is the deviation (in terms of norm = $\int [...]^2\,dx$) of the scanning function from a narrow function centered around x_i, such as the delta function $\delta(x - x_i)$. To avoid difficulties associated with the non-analytic nature of δ, it is easier to obtain a result for a narrow continuous function and then proceed to the limit. Although Backus and Gilbert (1970) preferred the spread measure $\int (x - x_i)^2\, s_i^2(x)\,dx$, we will nevertheless obtain a result for a measure $\int [s_i(x) - \delta(x - x_i)]^2\,dx$, because it provides a direct connection between the Gilbert-Backus solution and that obtained by the constrained linear inversion methods of Chapter 6.

The problem can be restated in terms of the approximation (in a least squares sense) of $\delta(x - x_i)$ by a linear combination of kernels, and we will derive the result in this way rather than more formally by finding an appropriate S for insertion into equation [7.9a].

Let $\phi(x - x_i)$ be a function which is integrable and square integrable on the interval with which we are concerned. To obtain for ϕ the least squares approximation, $\Sigma_j\, b_{ij}\, K_j(x)$, i.e. that which minimizes $\int [\boldsymbol{b}_i^*\, \boldsymbol{k}(x) - \phi_i(x)]^2\,dx$, one need only solve the Gauss normal equations for $i = 1, 2$, etc. For $i = 1$, these are:

$$
\left.\begin{aligned}
&b_{11} \int K_1^2(x)\,dx + b_{12} \int K_1(x)\,K_2(x)\,dx + \ldots + b_{1M} \int K_1(x)\,K_M(x)\,dx \\
&\qquad\qquad\qquad\qquad\qquad\qquad = \int K_1(x)\,\phi(x - x_1)\,dx \\
&b_{11} \int K_1(x)\,K_2(x)\,dx + b_{12} \int K_2^2(x)\,dx + \ldots \qquad = \int K_2(x)\,\phi(x - x_1)\,dx \\
&\cdot \\
&\cdot \\
&b_{11} \int K_1(x)\,K_M(x)\,dx + b_{12} \int K_2(x)\,K_M(x)\,dx + \ldots = \int K_M(x)\,\phi(x - x_1)\,dx
\end{aligned}\right\}
$$

The solution of the system of equations is:

$$
\begin{bmatrix} b_{11} \\ b_{12} \\ b_{13} \\ \cdot \\ \cdot \\ b_{1M} \end{bmatrix} = \mathbf{C}^{-1}\boldsymbol{y}_1 \qquad [7.11]
$$

where $\mathbf{C}$ is the covariance matrix $\|\int K_i(x)\, K_j(x)\, dx\|$ and we have written $\boldsymbol{y}_1$ to denote the vector $\int \boldsymbol{k}(x)\, \phi(x - x_1)\, dx$. For each $i = 1, 2, ..., n$, a different set of coefficients is obtained; these can be viewed as being successive rows of a matrix $\mathbf{B}$. In each instance the left-hand system of coefficients (i.e. the matrix $\mathbf{C}$) is the same; only the right-hand side of [7.11] changes, becoming successively $\boldsymbol{y}_2 = \int \boldsymbol{k}(x)\, \phi(x - x_2)\, dx$, $\boldsymbol{y}_3 = \int \boldsymbol{k}(x)\, \phi(x - x_3)\, dx$, and so on. Vectors:

$$\begin{bmatrix} b_{11} \\ b_{12} \\ b_{13} \\ \cdot \\ \cdot \\ b_{1M} \end{bmatrix} \quad \begin{bmatrix} b_{21} \\ b_{22} \\ b_{23} \\ \cdot \\ \cdot \\ b_{2M} \end{bmatrix} \quad \cdots \quad \begin{bmatrix} b_{M1} \\ b_{M2} \\ b_{M3} \\ \cdot \\ \cdot \\ b_{MM} \end{bmatrix}$$

are obtained by using $\boldsymbol{y}_1, \boldsymbol{y}_2, ..., \boldsymbol{y}$ in turn as the right-hand side of [7.11]. In matrix notation the entire sequence of operations is given as:

$$\mathbf{B}^* = \mathbf{C}^{-1}\mathbf{Y}$$

($\mathbf{B}^*$ because the vectors written above are columns of $\mathbf{B}^*$ rather than of $\mathbf{B}$). $\mathbf{Y}$ contains in its successive columns $\boldsymbol{y}_1, \boldsymbol{y}_2$, etc. This equation can be transposed to:

$$\mathbf{B} = \mathbf{Y}^*\mathbf{C}^{-1} \qquad [7.12]$$

$\mathbf{C}$ being symmetric and so unaffected by transposition.

In these formulae $\mathbf{Y}$ represents in essence some kind of tabulation of the kernel functions. This is quite apparent when ϕ is the limiting case, the delta function, for then the $\boldsymbol{y}_i$ become $\int \boldsymbol{k}(x)\, \delta(x - x_i)\, dx$, i.e. $\boldsymbol{k}(x_i)$, and the matrix $\mathbf{Y}$ becomes $\|K_i(x_j)\|$, or:

$$\begin{bmatrix} K_1(x_1) & K_1(x_2) & \cdots \\ K_2(x_1) & K_2(x_2) & \cdots \\ K_3(x_1) & K_3(x_2) & \cdots \\ \cdot & \cdot & \\ \cdot & \cdot & \\ K_M(x_1) & K_M(x_2) & \cdots \end{bmatrix}$$

a discrete $(M \times N)$ tabulation of the M kernels at tabular x-values $x_1, x_2, ..., x_N$. When $\phi(x - x_i)$ is not a delta function it still gives a locally averaged value of $\boldsymbol{k}(x)$ around the point $x = x_i$ and $\mathbf{Y}$ can be written $\|\overline{K_i(x_j)}\|$.

We now have for the spread measure $\int [s_i(x) - \delta(x - x_i)]^2 \, dx$ a formula for synthesizing the optimum $s_i(x)$ by a linear combination of kernels. This of course immediately provides an inversion formula since:

$$f(x_i) = \int \delta(x - x_i) \, f(x) \, dx$$

$$\cong \int \sum_j b_{ij} K_j(x) \, f(x) \, dx$$

or, in matrix-vector notation, using $\int K_j(x) \, f(x) \, dx = g_j$:

$$f(x) \cong \mathbf{B} g$$

or, substituting for **B**:

$$f(x) \cong \mathbf{Y}^* \mathbf{C}^{-1} g \qquad [7.13a]$$

or, writing out the vectors involved:

$$\begin{bmatrix} f(x_1) \\ f(x_2) \\ \cdot \\ \cdot \\ f(x) \end{bmatrix} \cong \mathbf{Y}^* \mathbf{C}^{-1} \int \begin{bmatrix} K_1(x) \\ K_2(x) \\ \cdot \\ \cdot \\ K(x) \end{bmatrix} f(x) \, dx \qquad [7.13b]$$

These equations are evidently very close to direct inversion formulae of the form:

$$f = \mathrm{A}^{-1} g$$

Indeed if we write out the various matrices in [7.13a], we have:

$$\mathbf{Y}^* = \begin{bmatrix} K_1(x_1) & K_2(x_1) & \ldots \\ K_1(x_2) & K_2(x_2) & \ldots \\ \cdot & \cdot & \\ \cdot & \cdot & \end{bmatrix}$$

$$\mathbf{C} = \begin{bmatrix} \int K_1^2(x) \, dx & \int K_1(x) \, K_2(x) \, dx & \ldots & \\ \int K_1(x) \, K_2(x) \, dx & \int K_2^2(x) \, dx & \ldots & \\ \cdot & \cdot & & \\ \cdot & \cdot & \ldots & \\ \int K_1(x) \, K_M(x) \, dx & \cdot & \ldots & \int K_M^2(x) \, dx \end{bmatrix}$$

The constrained inversion techniques of Chapter 6 utilized quadrature formulae to convert integrals of the form $\int K_i(x)\, f(x)\, dx$ into finite sums $\Sigma_j\, a_{ij}\, f(x_j)$. The elements of **C**, being integrals of this form, can be approximated by similar quadratures. The quadrature matrix **A** on multiplying the matrix $\mathbf{Y}^*$ gives a product in which the (i, j) element is:

$$a_{i1}K_j(x_1) + a_{i2}K_j(x_2) + \ldots$$

which is simply a quadrature approximation for $\int K_i(x)\, K_j(x)\, dx$, i.e. for the (i, j) element of the covariance matrix **C**. Thus to within the accuracy of the quadrature formula **C** can be written as $\mathbf{AY}^*$. Making that substitution in [7.13a] one obtains:

$$f(x_i) \cong \mathbf{Y}^*(\mathbf{Y}^*)^{-1}\mathbf{A}^{-1}\boldsymbol{g}$$

or:

$$f \cong \mathbf{A}^{-1}\boldsymbol{g}$$

which is simply the direct inversion formula obtained by inverting the quadrature, and already discussed in Chapter 6 (and shown to be found wanting). To stabilize the process a smoothing factor γ must be included to prevent the synthesizing coefficients (the elements of $\boldsymbol{r}$ in equation [7.9a]) from becoming excessively large. When such a constraint is included in the case just discussed (spread measured as distance from the delta function), the equation for the coefficients (7.11) is unchanged except that $\mathbf{C}^{-1}$ must be replaced by $(\mathbf{C}^*\mathbf{C} + \gamma\mathbf{I})^{-1}\mathbf{C}^*$. If the same γ is used for each set of coefficients,

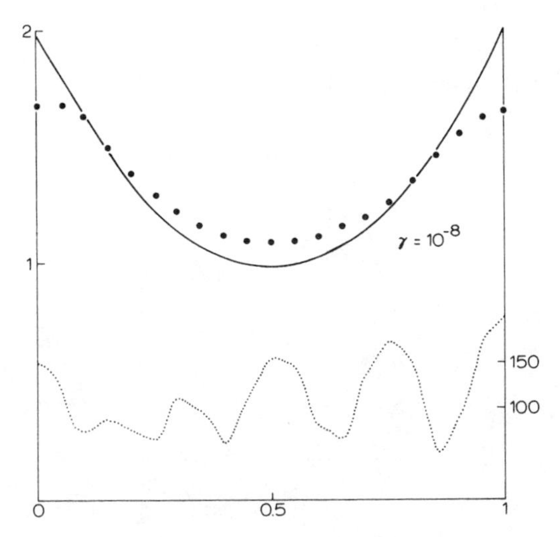

Fig. 7.4. Inversion of the standard problem by the Gilbert-Backus method using a value for γ which gave error magnifications shown on the lower (dotted) curve.

the inversion obtained for f becomes:

$$f \cong Y^*(C^*C + \gamma I)^{-1} C^*g$$

and using the quadrature $C = AY^*$, this becomes:

$$f \cong Y^*(YA^*AY^* + \gamma I)^{-1} YA^*g = [Y^{-1}(YA^*AY^* + \gamma I) Y^{*-1}]^{-1} A^*g$$

or:

$$f = (A^*A + \gamma I)^{-1} A^*g$$

which is simply the constrained linear inversion formula of Chapter 6 with a constraint matrix **H**. Before leaving the Gilbert-Backus and similar synthetic formulae, the result will be given of applying formulae [7.9a] and [7.10] to the standard inversion problem which has been used elsewhere. We have already (Fig. 7.2) shown the scanning functions obtained for these kernels ($x \exp(-y_i x)$, $y = (0.05)^{-1}, (0.1)^{-1}, ..., 1$) using the Gilbert-Backus spread measure, and Fig. 7.4 shows the result of inversion of the standard problem by the Gilbert-Backus method.

The results are not greatly different from those of the methods of Chapter 6. Indeed the methods fundamentally differ in only one main point: constraints are applied to f' in the case of constrained inversion, but in synthesis methods the constraints are applied to coefficients in $\mathbf{r}$; in the latter case each $|\mathbf{r}_i|^2$ is constrained, in the former it is $\Sigma_i\ |\mathbf{r}_i^* \mathbf{g}|^2$ or some related norm. In both situations a bound is placed on a norm associated with the array of numbers r_{ij}.

7.3. SOLUTION IN TERMS OF KERNELS

From M measurements $\int K_1(x) f(x)\, dx$, $\int K_2(x) f(x)\, dx$, ..., $\int K_M(x) f(x)\, dx$ one can evidently only obtain information on the projection of $f(x)$ on to the function subspace spanned by the kernels $K_1(x), K_2(x), ..., K_M(x)$. In other words, the most general solution which can be legitimately obtained is:

$$f(x) = \xi_1 K_1(x) + \xi_2 K_2(x) + ... + \xi_M K_M(x) + \psi(x)$$

where $\psi(x)$ is any (arbitrarily large) function orthogonal to the set of functions $K_1(x), K_2(x), ..., K_M(x)$. Making this substitution into the fundamental integral equation one gets:

$$g_i = \int \sum_j K_i(x)\, K_j(x)\, \xi_j dx$$

or $\mathbf{g} = \mathbf{C}\boldsymbol{\xi}$

C being the covariance matrix $||\int K_i(x)\, K_j(x)\, dx||$. We have already seen that **C** is highly ill-conditioned (possesses very small eigenvalues) and the above equation is no more amenable to direct inversion than those encountered earlier. However, it does have the advantage that components orthogonal to the kernels are explicity excluded from the solution, whereas in other methods they may tend to creep in. However, further smoothing of the solution by some means is still necessary in most instances and this is now less easy since explicit reference is no longer made to $f(x)$ or $\boldsymbol{f}$ in the equation $\boldsymbol{g} = \mathbf{C}\xi$.

Since it is those components in ξ which are associated with small eigenvalues that are inadequately represented in the measurements and that give rise to instability, deletion of such components is a meaningful and desirable smoothing procedure. Let ξ be expanded in terms of the eigenvectors of $\mathbf{C}$:

$$\xi = \beta_1 \boldsymbol{u}_1 + \beta_2 \boldsymbol{u}_2 + \ldots + \beta_M \boldsymbol{u}_M$$

so:

$$\boldsymbol{g} = \lambda_1 \beta_1 \boldsymbol{u}_1 + \lambda_2 \beta_2 \boldsymbol{u}_2 + \ldots + \lambda_M \beta_M \boldsymbol{u}_M$$

and:

$$\beta_k = \lambda_k^{-1} (\boldsymbol{u}_k \cdot \boldsymbol{g})$$

If we agree to delete components associated with eigenvalues less than some prescribed value e, the corresponding β's are simply set to zero or just dropped from the summation.

The solution thus obtained can be written in matrix form if a diagonal matrix $\mathbf{D}(e)$ is derived from Λ^{-1} by deleting from Λ^{-1} all reciprocal eigenvalues greater than e^{-1}. Since $\mathbf{C}$ is symmetric, real and positive, there are no complex or negative eigenvalues.

The unfiltered inverse of $\boldsymbol{g} = \mathbf{C}\xi$ can readily be written in terms of $\mathbf{U}$ and Λ ($\mathbf{C}$ being equal to $\mathbf{U}\Lambda\mathbf{U}^*$); it is:

$$\left.\begin{aligned} \xi &= \mathbf{U}\Lambda^{-1}\mathbf{U}^* \boldsymbol{g} \\ f(x) &= \xi^* \boldsymbol{k}(x) \end{aligned}\right\}$$

The coefficients $\beta_1, \beta_2, \ldots$, in the expansion $\xi = \beta_1 \boldsymbol{u}_1 + \beta_2 \boldsymbol{u}_2 \ldots$, are given as a vector β by $\xi = \mathbf{U}\beta$, so β is $\Lambda^{-1}\mathbf{U}^*\boldsymbol{g}$. To filter the solution by removing elements of β associated with eigenvalues less than e, Λ^{-1} is simply replaced by $\mathbf{D}(e)$, so (using primes to denote filtered vectors or functions):

$$\beta' = \mathbf{D}\mathbf{U}^* \boldsymbol{g}$$

$$\xi' = \mathbf{U}\mathbf{D}\mathbf{U}^* \boldsymbol{g}$$

$$f'(x) = (\xi')^* \boldsymbol{k}(x) = \boldsymbol{g}^* \mathbf{U}\mathbf{D}\mathbf{U}^* \boldsymbol{k}(x) \qquad [7.14]$$

To realize this result an eigenanalysis of $\mathbf{C}$ must therefore be carried out. Formula [7.14] essentially replaces $\mathbf{C}$ by a matrix with the same eigenvectors but with infinitely large eigenvalues where those of $\mathbf{C}$ were $<e$. In prac-

tice this procedure is rarely any better than the simpler procedure of using a form of equation [6.5]; i.e. by applying a constrained inversion to $\mathbf{C}\xi = g$, using as a constraining quadratic the square norm of ξ, so that:

$$\xi' = (\mathbf{C}^*\mathbf{C} + \gamma\mathbf{I})^{-1}\,\mathbf{C}^*\boldsymbol{g}$$
$$f'(x) = \boldsymbol{g}^*\mathbf{C}(\mathbf{C}^*\mathbf{C} + \gamma\mathbf{I})^{-1}\,\boldsymbol{k}(x) \qquad [7.15]$$

In terms of the eigenvalues of $\mathbf{C}^*\mathbf{C}$, this formula leaves components with $\lambda >> \gamma$ effectively unchanged while components associated with eigenvalues much less than γ are effectively deleted. If:

$$\mathbf{C}^*\mathbf{C}\boldsymbol{u}_k = \lambda_k \boldsymbol{u}_k$$

then:

$$(\mathbf{C}^*\mathbf{C} + \gamma\mathbf{I})\boldsymbol{u}_k = (\lambda_k + \gamma)\boldsymbol{u}_k$$

so if:

$$\xi = \beta_1\boldsymbol{u}_1 + \beta_2\boldsymbol{u}_2 + \ldots + \beta_M\boldsymbol{u}_M$$
$$\mathbf{C}^*\boldsymbol{g} = \mathbf{C}^*\mathbf{C}\xi = \lambda_1\beta_1\boldsymbol{u}_1 + \lambda_2\beta_2\boldsymbol{u}_2 + \ldots + \lambda_M\beta_M\boldsymbol{u}_M$$
$$(\mathbf{C}^*\mathbf{C} + \gamma\mathbf{I})^{-1}\,\mathbf{C}^*\boldsymbol{g} = \frac{\lambda_1}{\lambda_1 + \gamma}\beta_1\boldsymbol{u}_1 + \frac{\lambda_2}{\lambda_2 + \gamma}\beta_2\boldsymbol{u}_2 + \ldots + \frac{\lambda_M}{\lambda_M + \gamma}\beta_M\boldsymbol{u}_M$$

The filtering effect of [7.15] is therefore more gradual, whereas equation [7.14] is equivalent to a discontinuous filtering where the solution:

$$\beta_1\boldsymbol{u}_1 + \beta_2\boldsymbol{u}_2 + \ldots + \beta_M\boldsymbol{u}_M$$

is abruptly terminated when $\lambda_n < e$. Fig. 7.5 illustrates the different filterings achieved by these two formulae.

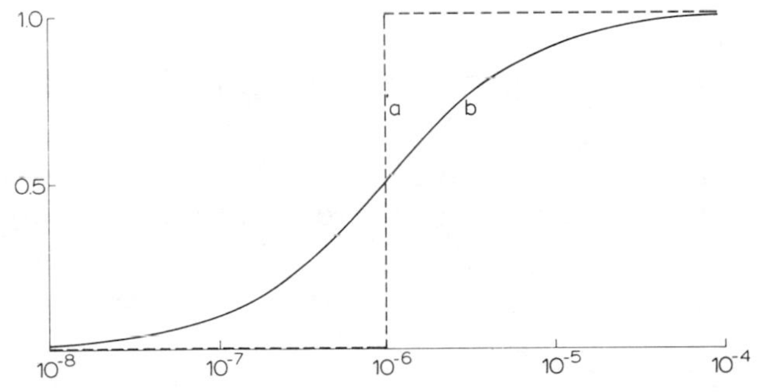

Fig. 7.5. Comparison of filtering by two methods discussed in the text: equations [7.14] and [7.15] respectively apply for curves marked *a* and *b*.

7.4. THE PRONY ALGORITHM — A NON-LINEAR INVERSION METHOD

An algebraic problem which has all the features of inversion problems, and into which some temperature sensing problems can be transformed, is that of fitting a set of exponentials to a given function or a numerical set of data. This involves finding both a's and α's in an expression of the form:

$$a_1 \, e^{\alpha_1 x} + a_2 \, e^{\alpha_2 x} + \dots + a_n \, e^{\alpha_n x}$$

so as to make that expression take prescribed values, say $y_1, y_2, \dots, y_n$ at certain values of x, $x_1, x_2, \dots, x_n$. This is a fundamentally non-linear problem — the problem of finding the a's when the α's are unknowns.

The procedure for solving this problem was given by Prony and is a very neat and appealing method.

Let:

$$\left.\begin{aligned} y_1 &= a_1 \, e^{\alpha_1 \xi} + a_2 \, e^{\alpha_2 \xi} + \dots + a_n \, e^{\alpha_n \xi} \\ y_2 &= a_1 \, e^{2\alpha_1 \xi} + a_2 \, e^{2\alpha_2 \xi} + \dots + a_n \, e^{2\alpha_n \xi} \\ &\cdot \\ &\cdot \end{aligned}\right\}$$

represent the values of y at the uniformly spaced tabular values $x = \xi, 2\xi, 3\xi, \dots, 2n\xi$. The number of unknowns is $2n$, consisting of n unknown coefficients $a_1, a_2, \dots, a_n$ and n exponential factors $\alpha_1, \alpha_2, \dots, \alpha_n$; Since $e^{k\alpha\xi} = (e^{\alpha\xi})^k$, a change of variable from the α's to $u_1 = e^{\alpha_1 \xi}$, $u_2 = e^{\alpha_2 \xi}$, etc., leads to the equivalent and somewhat simpler system of equations:

$$\left.\begin{aligned} y_1 &= a_1 u_1 + a_2 u_2 + \dots + a_n u_n \\ y_2 &= a_1 u_1^2 + a_2 u_2^2 + \dots + a_n u_n^2 \\ &\cdot \\ &\cdot \\ y_{2n} &= a_1 u_1^{2n} + a_2 u_2^{2n} + \dots + a_n u_n^{2n} \end{aligned}\right\}$$

these equations can be multiplied by arbitrary factors and added to obtain new but equivalent equations. The choice of these factors is the crucial point in the algorithm; they are chosen as follows: let the first N equations be multiplied by $\beta_1, \beta_2, \dots, \beta_n$ respectively and added to the $(n+1)$th equation, giving:

$$y_{n+1} + \sum_1^n \beta_j y_j = a_1 \{u_1^{n+1} + \sum_1^n \beta_j u_1^j\} + a_2 \{u_2^{n+1} + \sum_1^n \beta_j u_2^j\} + \dots$$

The factors multiplying the a's on the right side are respectively polynomials in u_1, in u_2, and so on; by *defining* the β's as the coefficients of a polynomial $\Sigma_j\ \beta_j u^j + u^{n+1}$ which has roots $u_1, u_2, \ldots, u_n$, the right side of the above equation can be made to vanish. If now the same value of the β's are used on the second, third, ..., $(n+1)$th equation and the results added to the $(n+2)$th equation we obtain:

$$y_{n+2} + \sum_1^n \beta_j y_{j+1} = a_1\{u_1^{n+2} + \sum_1^n \beta_j u_1^{j+1}\} + a_2\{u_2^{n+2} + \sum_1^n \beta_j u_2^{j+1}\} + \ldots$$

but if $u^{n+1} + \Sigma_1^n\ \beta_j u^j$ vanishes, so also does $u^{n+2} + \Sigma_j\ \beta_j u^{j+1} = u\{u^{n+1} + \Sigma_1^n\ \beta_j u^j\}$ — thus if we repeat the procedure for all successive sets of $n+1$ equations, we obtain in every case zero on the right side. The unknowns $u_1, u_2, \ldots, u_n$ in this way have been eliminated and we have a simple linear system of equations in which the y's are known and the β's unknown:

$$\left.\begin{array}{l} \beta_1 y_1 + \beta_2 y_2 + \ldots + \beta_n y_n = -y_{n+1} \\ \beta_1 y_2 + \beta_2 y_3 + \ldots + \beta_n y_{n+1} = -y_{n+2} \\ \cdot \\ \cdot \\ \beta_1 y_n + \beta_2 y_{n+1} + \ldots + \beta_n y_{2n-1} = -y_{2n} \end{array}\right\}$$

Solving these equations using the given y-values, one obtains the β's explicitly. The unknowns u_1, u_2, etc., can then be obtained since they are by definition the roots of the polynomial $u^{n+1} + \Sigma_1^n\ \beta_j u^j$. Finally the system of equations relating the y's, u's, and a's can be solved for the a's now that the u's have been determined. In summary therefore the Prony algorithm consists of:

(1) solving the linear system:

$$\left.\begin{array}{l} \beta_1 y_1 + \beta_2 y_2 + \ldots + \beta_n y_n = -y_{n+1} \\ \beta_1 y_2 + \beta_2 y_3 + \ldots + \beta_n y_{n+1} = -y_{n+2} \\ \cdot \quad \cdot \qquad \cdot \quad \cdot \\ \cdot \quad \cdot \qquad \cdot \quad \cdot \\ \beta_1 y_n + \beta_2 y_{n+1} + \ldots + \beta_n y_{2n-1} = -y_{2n} \end{array}\right\}$$

to obtain the array β;

(2) finding the roots of the polynomial:

$$u^{n+1} + \sum_{j=1}^n \beta_j u^j, \qquad \text{(i.e. of } \beta_1 + \beta_2 u + \ldots + \beta_n u^{n-1} + u^n)$$

to obtain $u_1, u_2, \ldots u_n$;

(3) solving the linear system:

$$\left.\begin{aligned} y_1 &= a_1 u_1 + a_2 u_2 + \ldots + a_n u_n \\ y_2 &= a_1 u_1^2 + a_2 u_2^2 + \ldots + a_n u_n^2 \\ &\cdot \\ &\cdot \\ y_n &= a_1 u_1^n + a_2 u_2^n + \ldots + a_n u_n^n \end{aligned}\right\}$$

for the array a:

(4) if the α's are explicitly required they are simply:

$$\xi^{-1} \ln u_k \qquad (k = 1, 2, \ldots n)$$

Application to inversion problems involving the Laplace transform

Many inversion problems can be expressed as a Laplace transform inversion. The radiation emerging from a horizontally homogeneous layer emitting and absorbing radiation according to Planck's and Kirchhoff's laws varies with angle of emergence $\cos^{-1} \mu$ (measured from the vertical) according to the following relationship:

$$I(\mu) = \int_0^\infty B(\tau)\, \mathrm{e}^{-\tau/\mu}\, \mathrm{d}\tau/\mu$$

or if $\mu^{-1} = y$:

$$y^{-1} I(y) = \int_0^\infty B(\tau)\, \mathrm{e}^{-y\tau}\, \mathrm{d}\tau \qquad [7.16]$$

Here $B(\tau)$ is the Planck radiance at a normal optical thickness τ from the observing instrument (B determines temperature uniquely once the wavelength or frequency is specified). By measuring the radiance at different angles of emergence, one essentially obtains $g(y)$ at a succession of values of y; these can be spaced uniformly if μ is suitably spaced. If $L\{f\}$ denotes the Laplace transform of $f(x)$, i.e. $\int_0^\infty \mathrm{e}^{-yx} f(x)\, \mathrm{d}x$, then [7.9a] implies that the emerging radiation, multiplied by μ and viewed as a function of y, the reciprocal of μ, gives the Laplace transform of the distribution of B with τ (and hence temperature with height).

King (1964) has applied the Prony algorithm to the inversion of this problem by envisaging the layer to be approximated by a number of slabs each of which has a constant gradient of B versus τ — which is almost equivalent to a constant temperature gradient within each slab. We have, integrating by parts:

$$\int_0^\infty B(\tau)\,\mathrm{e}^{-y\tau}\,\mathrm{d}\tau = \frac{B(\tau)\,\mathrm{e}^{-y\tau}}{y}\Bigg|_0^\infty + \frac{1}{y}\int_0^\infty \frac{\mathrm{d}B}{\mathrm{d}\tau}\,\mathrm{e}^{-y\tau}\,\mathrm{d}\tau$$

$$= \frac{B_0}{y} + \frac{1}{y}\int_0^\infty \frac{\mathrm{d}B}{\mathrm{d}\tau}\,\mathrm{e}^{-y\tau}\,\mathrm{d}\tau$$

Hence:

$$I(y) - B_0 = \int_0^\infty \frac{\mathrm{d}B}{\mathrm{d}\tau}\,\mathrm{e}^{-y\tau}\,\mathrm{d}\tau$$

or if ΔB_k denotes the value of $\mathrm{d}B/\mathrm{d}\tau$ within the kth slab, then summing over all slabs one obtains:

$$I(y) - B_0 \quad = \sum_k \Delta B_k \int_{\tau_k}^{\tau_{k+1}} \mathrm{e}^{-y\tau}\,\mathrm{d}\tau$$

$$= \sum_k \Delta B_k \int_{\tau_k}^{\tau_{k+1}} \mathrm{e}^{-y\tau}\,\mathrm{d}\tau$$

$$yI(y) - yB_0 = \sum_k \Delta B_k\,\mathrm{e}^{-y\tau_k}\,(1 - \mathrm{e}^{-y\Delta\tau_k}) = \sum_k \Delta B'_k\,\mathrm{e}^{-y\tau_k} \qquad [7.17]$$

where the kth slab extends from τ_k to τ_{k+1}. If this is defined in advance a linear problem in the unknows $\Delta B'_k$ results, but if the τ-intervals are not fixed a priori, equation [7.17] can still be solved by means of the Prony algorithm. King has demonstrated that allowing the slab thickness to be a variable in this way can produce a considerable improvement in temperature profiles retrieved from radiation measurements.

It is important, however, to point out that there is no inherent stability in the algorithm — indeed Lanczos (1956) gives the following example of its "application". Starting out with the hypothesis:

$$y = 0.0951\,\mathrm{e}^{-x} + 0.8607\,\mathrm{e}^{-3x} + 1.5576\,\mathrm{e}^{-5x}$$

y was calculated at 24 values of x from a to 24 and the Prony algorithm applied. To within the three significant figures that were carried the result:

$$y = 2.202\,\mathrm{e}^{-4.45x} + 0.305\,\mathrm{e}^{-1.58x}$$

was obtained and found to give y-values which were acceptably close to the original data! This is of course merely another manifestation of interdependence among the smooth, indeed monotonic, exponential functions. The reader will find it informative to compute and plot out some values of the two combinations of exponentials given above.

7.5. LANDWEBER ITERATION

When the kernel and $g(y)$ are defined analytically, an iterative process introduced by Landweber (1951) for the equation $g(y) = \int K(y, x) f(x)\, dx$ leads to convergence to the true solution provided the eigenvalues of a certain operator (closely allied to $\mathbf{A}^*\mathbf{A}$) lie between 0 and 2. The iteration proceeds from the nth iterate $f_n(x)$ to the $(n + 1)$th as follows:

(1) compute the residual $g(y) - \int K(y, x) f_n(x)\, dx = r_n(y)$

(2) compute $f_{n+1}(x) = f_n(x) + \int K(x, y)\, r_n(y)\, dy$

Note that the integral operation in (2) is with respect to y, and the arguments of K are reversed as compared to (1).

Real inversion problems rarely provide analytically defined kernels and g's, but the above iteration scheme has an obvious discrete analogue which applies to the inversion of $\mathbf{A}\boldsymbol{f} = \boldsymbol{g}$:

(1) compute the residual $\boldsymbol{g} - \mathbf{A}\boldsymbol{f}_n = \boldsymbol{r}_n$

(2) compute $\boldsymbol{f}_{n+1} = \boldsymbol{f}_n + \mathbf{A}^*\boldsymbol{r}_n$

The eventual result of a sequence of such iterations is therefore:

$$\begin{aligned} f_n &= f_{n-1} + \mathbf{A}^*(g - \mathbf{A}f_{n-1}) = \mathbf{A}^*g + (\mathbf{I} - \mathbf{A}^*\mathbf{A})f_{n-1} \\ &= \mathbf{A}^*g + (\mathbf{I} - \mathbf{A}^*\mathbf{A})[\mathbf{A}^*g + (\mathbf{I} - \mathbf{A}^*\mathbf{A})f_{n-2}] \\ &= \ldots = \ldots = \ldots \\ &= \{\mathbf{A}^* + (\mathbf{I} - \mathbf{A}^*\mathbf{A})\,\mathbf{A}^* + (\mathbf{I} - \mathbf{A}^*\mathbf{A})^2\,\mathbf{A}^* \ldots\}\, g + (\mathbf{I} - \mathbf{A}^*\mathbf{A})^n f_0 \end{aligned} \qquad [7.18]$$

The bracketed matrix series is simply a series of the form $1 + \chi + \chi^2 \ldots$, which has the sum to infinity $1/(1 - \chi)$ when $|\chi| < 1$, and similarly the matrix series can be written $[\mathbf{I} - (\mathbf{I} - \mathbf{A}^*\mathbf{A})]^{-1}\mathbf{A}^* = (\mathbf{A}^*\mathbf{A})^{-1}\mathbf{A}^*$ provided the eigenvalues of $(\mathbf{I} - \mathbf{A}^*\mathbf{A})$ are numerically less than unity. As n increases the expression $(I - \mathbf{A}^*\mathbf{A})f_0$ diminishes at least as fast as $(1 - \lambda_{max})^n$ where λ_{max} is the largest (numerically) of the eigenvalues of $\mathbf{A}^*\mathbf{A}$. Thus as $n \to \infty$ the above iteration tends to:

$$f_\infty = (\mathbf{A}^*\mathbf{A})^{-1}\,\mathbf{A}^*g$$

which is the least squares solution (f'' of equation [6.2]). In the limit, therefore, this kind of iteration possesses no more stability than the least squares solution, which already has been demonstrated to be as bad or even worse than direct inversion. The method, however, possesses a measure of stability if the iteration can be stopped fairly early. The solution f after n iterations can evidently be written:

$$f_n = [\mathrm{I} - (\mathrm{I} - \mathbf{A}^*\mathbf{A})]^{-1} [\mathrm{I} - (\mathrm{I} - \mathbf{A}^*\mathbf{A})^{n+1}] \mathbf{A}^* g + (\mathrm{I} - \mathbf{A}^*\mathbf{A})^n f_0$$

(summing the matrix series in the same way that a scalar geometric series is summed). If these operations are looked at from the point of view of an eigenvector of $\mathbf{A}^*\mathbf{A}$, the first term is equivalent to multiplication by:

$$[1 - (1 - \lambda)]^{-1} [1 - (1 - \lambda)^{n+1}] = \frac{1}{\lambda} [1 - (1 - \lambda)^{n+1}]$$

for $|\lambda| < 1$ this tends to λ^{-1} as $n \to \infty$, but for λ close to 1 the approach to λ^{-1} is much faster than for small λ. For example, if $\lambda = 0.9$, $\lambda^{-1} = 1.1111$ and we have at $n = 10$ the value $1.1111 - 1.11111 \times 10^{-11}$, but for $\lambda = 0.01$, $\lambda^{-1} = 100$, the value at $n = 10$ is $100(1 - 0.895) = 10.47$. Thus if the iteration is stopped after n iterations there is a filtering of the eigen vector contribution in proportion to $1 - (1 - \lambda)^{n+1}$. Now, by a well-known algebraic relationship:

$$e^{-x} = \lim_{n \to \infty} \left(1 - \frac{x}{n}\right)^n$$

Thus $(1 - \lambda)^{n+1}$ tends to $e^{-(n+1)\lambda}$ and is in fact well approximated by $e^{-(n+1)\lambda}$ for quite moderate values of n except when λ is very close to unity. The application of n cycles of Landweber iteration is therefore equivalent to a direct inversion by equation [6.2] preceded by a filtering of $\mathbf{A}^* g$ in which the projection of $\mathbf{A}^* g$ along a given eigenvector $\boldsymbol{u}$ of $\mathbf{A}^*\mathbf{A}$ is reduced in proportion to a filtering factor which is close to unity when λ is near unity and for smaller values of λ is exponential in character, being in fact $1 - e^{-(n+1)\lambda}$. A graph of this factor is shown in Fig. 7.6 for $n = 10, 100$, and 1000 cycles of iteration.

The filtering aspect of constrained inversion by equation [6.3] has already been examined; it was seen to be described by a filter factor $\lambda/(\lambda + \gamma)$ when a Lagragian multiplier γ and $\mathbf{H} = \mathbf{I}$ was used. This factor for $\gamma = 0.1, 0.01$ and 0.001 has been plotted (dotted curve) on the figure. It is apparent that the application of n cycles of iteration is algebraically very similar indeed to use of equation [6.3] with $\gamma = n^{-1}$. The second term in f above was $(\mathrm{I} - \mathbf{A}^*\mathbf{A})^n f_0$. Since f_0 is an arbitrary first guess, this could be taken as zero and the second term eliminated. Alternately the problem of obtaining f starting from f_0 can be rephrased as that of obtaining $(f - f_0)$ as the unknown, starting from zero. Thus in place of the first iteration:

$$f_1 = f_0 + \mathbf{A}^* r_0; \qquad r_0 = g - \mathbf{A} f_0$$

One can write the equivalent statement:

$$(f_1 - f_0) = 0 + \mathbf{A}^* r_0'; \qquad r_0' = g' - \mathbf{A} \cdot 0$$

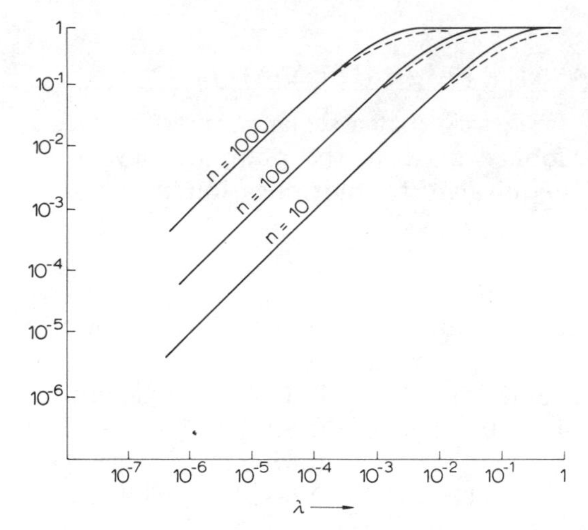

Fig. 7.6. Filtering effect of Landweber iteration compared with that given by constrained linear inversion by equation [6.4] (with $\mathbf{H} = \mathbf{I}$).

with:

$$g' = g - \mathbf{A} f_0$$

f_0 therefore plays a purely incidental role in the algorithm and, as we have seen, the term containing it diminishes rapidly in any event.

As a matter of interest it may be noted that the above iteration scheme is very closely related to a scheme whereby an approximation to $\mathbf{A}^{-1}$ can be refined. If $\mathbf{B}$ is an approximation to $\mathbf{A}^{-1}$, and $\mathbf{X} = \mathbf{I} - \mathbf{BA}$ (which would vanish if $\mathbf{B}$ were exactly $\mathbf{A}^{-1}$), then if we calculate:

$$(\mathbf{I} + \mathbf{X} + \mathbf{X}^2 + \dots)\, \mathbf{B}$$

The series, if it converges, gives $\mathbf{A}^{-1}$. To show this one need only note that if $\mathbf{S}_n$ is the sum of n terms of the series in parentheses, then:

$$(\mathbf{I} - \mathbf{X})\, \mathbf{S}_n = \mathbf{I} - \mathbf{X}^n$$

$$\mathbf{S}_n = (\mathbf{I} - \mathbf{X})^{-1} (\mathbf{I} - \mathbf{X}^n)$$

if $\mathbf{X}^n \to 0$ (which it will if all the eigenvalues of $\mathbf{X}$ are numerically less than one), then:

$$\lim_{n \to \infty} \mathbf{S}_n = (\mathbf{I} - \mathbf{X})^{-1}$$

and so, since $\mathbf{BA} = \mathbf{I} - \mathbf{X}$, one has:

$$\mathbf{A} = \mathbf{B}^{-1} (\mathbf{I} - \mathbf{X})$$

hence:

$$A^{-1} = (I - X)^{-1} B = (I + X + X^2 + ...) B$$

7.6. ITERATIVE, NON-LINEAR METHODS OF INVERSION

All methods of inversion described up to now are linear, in the sense that if a solution f_1 is given by a measured g_1 and a solution f_2 by a measured g_2 then $f_1 + f_2$ would be the solution given by a measured $g_1 + g_2$.

A repetitive procedure may start with an initial guess or trial solution f_1 and then adjust the solution additively by computing corrections or adjustments so that the final solution is $f_1 + \phi_1 + \phi_2 + ...$ However, such a procedure is still linear. On the other hand, if the initial solution (and subsequent improved solutions) are adjusted by *multiplication* of f_1 by a correcting or adjusting factor, so that the final solution is $f_1 \cdot \phi_1 \cdot \phi_2 ...$, such procedures are generally non-linear.

From what has been said in earlier chapters it is apparent that any successful method of inversion must implicitly or explicitly limit the solution and not allow oscillations to develop at frequencies to which the measurements lack sensitivity. The success of iterative inversion methods (linear or non-linear) must depend on the exclusion of these frequencies from the solution. Evidently the corrections ϕ_1, ϕ_2, etc., must be smooth. Iterative methods in which a first guess is altered by addition of a correcting term likewise are not restricted to non-negative functions, although at each stage of iteration any negative values can be removed. It is, however, possible to arrange *non-linear* iterative methods so that negative values can never be generated. If the iteration is written:

$$f_1(x) = \phi_1(x)\, f_0(x)\,,$$

$$f_2(x) = \phi_2(x)\, f_1(x)\,,$$

etc.

then if $f_0(x)$ is positive for all x in the interval and the adjusting ϕ's are also positive, the iterates will continue to be positive. This is an advantage of non-linear methods — they can be arranged so that the iterated solutions are at all stages positive. The constraints which were feasible in linear inversion methods did not in any case provide an explicit non-negativity constraint. Since large-amplitude high-frequency components must give negative values to the solution in some regions, a non-negative constraint is a powerful deterrent to the generation of high frequency oscillations.

If at any stage of iteration we have generated a function $f_n(x)$ which is an approximation of sorts to $f(x)$, then an obvious estimate of $f_n(x)$ as an approximation to $f(x)$ is provided by the residuals:

$$\eta_1 = \int K_1(x)\, f_n(x)\, dx - g_1$$

$$\eta_2 = \int K_2(x)\, f_n(x)\, dx - g_2$$

$$\vdots$$

$$\eta_M = \int K_M(x)\, f_n(x)\, dx - g_M$$

An exact solution would have $\eta_1 = \eta_2 = ..., = \eta_M = 0$ but in the presence of measurement errors one would terminate any search for a solution once each residual was less than or equal to the probable error of the corresponding measurement. If a given residual, say η_1, is negative it implies that $\int K_1(x)\, f_n(x)\, dx$ is too small and that $f_n(x)$ should be increased in the region where $K_1(x)$ does not vanish. To produce a given change in the residual η_1 the least increase in $f_n(x)$ will be needed if it is applied where $K_1(x)$ is greatest. If $f_n(x)$ and $K_1(x)$ are tabulated at $x = x_1, x_2, ..., x_N$, the integral $\int K_1(x_i)\, f_n(x_i)\, dx$ can be approximated by $S = \Sigma_k\, K_1(x_k)\, f_n(x_k)$. Since $\partial S/\partial f(x) = K_1(x)$, the change in $f_n(x)$ required to change S by ΔS is $\Delta S/K_1(x)$; the least overall change in the numbers $f(x_1)$, $f(x_2)$, ..., $f(x_N)$ is needed if the value of f is altered at that x where $K_1(x)$ is a maximum.

This is the basis of a procedure introduced by Chahine (1970) for iterative solution of inversion problems. Since $f(x)$ is only altered at values of x where the kernels attain maxima, one can only use tabular points which correspond to such maxima. If we order the kernels in the sequence of the x-values at their maxima, which we will call $x_1, x_2, ..., x_M$, then the inversion procedure can be summarized as follows when $f^n(x_j)$, $j = 1, 2, ..., N$, represents the iterated solution after n cycles of iteration:

(1) compute
$$y_1 = \int K_1(x)\, f_n(x)\, dx$$

$$y_2 = \int K_2(x)\, f_n(x)\, dx$$

$$\vdots$$

$$y_M = \int K_M(x)\, f_n(x)\, dx$$

using an appropriate quadrature formula.

(2) adjust $f_n(x_1)$ using y_1 (since it is postulated that $K_1(x)$ exhibits a maxi-

mum at x_1), to obtain a new value $f_{n+1}(x_1)$, such that:

$$f_{n+1}(x_1) = (g_1/y_1) f_n(x_1)$$

and proceed similarly through the tabular x-values, i.e.:

$$f_{n+1}(x_j) = (g_j/y_j) f_n(x_j) \qquad [7.19]$$

Note that g_j is positive if the physics of the problem requires it (i.e. if $f(x)$ and $K_j(x)$ are everywhere non-negative) and that y_j will be positive if $f_n(x)$ is positive. So if we start with a first guess $f_0(x)$ which is positive everywhere in the interval over which the integration is made, the subsequent iterates *cannot* become negative. A similar but not identical procedure introduced by Chahine is to evaluate the ratios $r_1 = g_1/y_1$, $r_2 = g_2/y_2$, etc., and adjust each $f_n(x_j)$ using all the r's, by combining them into weighted averages in which the weight of r_k is proportional to the contribution of the ordinate at x_j to the kth integral — which for equally spaced x is the $K_k(x_j)$, otherwise it is the corresponding quadrature coefficient a_{kj}. Thus:

$$f_{n+1}(x_j) = \hat{r}_j f_n(x_j)\ ; \qquad \hat{r}_j = \sum_i K_i(x_j)\, r_i / \sum_i K_i(x_j) \qquad [7.20]$$

A modification of the above procedure gives it more flexibility, allowing one to use tabular x's which do not necessarily coincide with the peak of any kernel, while at the same time making the adjustment to $f(x)$ in a smoother manner than that of equation [7.20]. If the number of kernels and tabular x's become large it is possible for the x_j to become too close in the sense that none of the kernels change appreciably over such a distance; it is then possible for, say, an upward adjustment of a certain $f(x_k)$ and a downward adjustment of the neighboring $f(x_{k+1})$ to introduce high-frequency components into the solution. There are clearly grounds for attempting to make the change in $f(x)$ spectrally similar to the kernels, and not more oscillatory. This is achieved if $f(x)$ is multiplied by a function $[1 + \xi K(x)]$; this will become negative only if the product of ξ and the maximum value of $K(x)$ can numerically exceed unity; assuming that the kernels are scaled so that $[K_i(x)]_{max} \leqslant 1$ — and that is always possible — then the correcting factor cannot become negative if $\xi \geqslant -1$. When $g_i > y_i$ some increase in $f(x)$ is called for and vice versa. The simplest choice for ξ_i is $(g_i/y_i - 1)$, which makes no change in $f(x)$ if $g_i = y_i$ and ranges between -1 and an indefinitely large positive number for the positive g, y, K, g and f with which one is normally concerned. Thus provided $|K_i(x)| < 1$ for all relevant x, the sequence of functions generated by the scheme:

$$f_{k+1}(x) = [1 + \xi_k K_i(x)]\, f_k(x) \qquad [7.21]$$

remain positive if the iteration is commenced with a positive $f_0(x)$. This iteration can be applied in a manner similar to the Chahine method. Determination of a solution by this iteration would proceed as follows, commencing with an initial $f_0(x)$:

(1) compute $y_1 = \int K_1(x)\, f_0(x)\, dx$

$$y_2 = \int K_2(x)\, f_0(x)\, dx$$

$$\cdot$$

$$\cdot$$

$$y_M = \int K_M(x)\, f_0(x)\, dx$$

(2) adjust to obtain the next iterate $f_1(x)$, using:

$$f_1(x) = [1 + \xi_1 K_1(x)][1 + \xi_2 K_2(x)] \ldots [1 + \xi_M K_M(x)]\, f_0(x)$$

where $1 + \xi_k = g_k / y_k$.

An alternative procedure makes the adjustment $[1 + \xi_1 K_1(x)]\ f_0(x)$ before calculating y_2, and so on, updating the iterate continuously rather than computing a complete set of y's for one $f(x)$. There does not appear to be any great difference between the two procedures. Since the methods are iterative one can also restrict the magnitude of each ξ to some interval, such as $0.1 > \xi > -0.1$, using a full computed value of ξ only when it lies within the specified limits.

When the ξ's are small, as they can be forced to be and which they become as an iteration proceeds, then to the extent that products of ξ's can be neglected, the result of iterating from a first guess $f_0(x)$ = constant is to produce a linear combination of kernels. In fact the end result of a sequence of iterations by equation [7.21] is:

$$\begin{aligned} f'(x) = \ & [1 + \xi_1 K_1(x)][1 + \xi_2 K_2(x)] + \ldots + [1 + \xi_M K_M(x)] \\ & \cdot [1 + \xi_1' K_1(x)][1 + \xi_2' K_2(x)] + \ldots + [1 + \xi_M' K_M(x)] \\ & \cdot [1 + \xi_1'' K_1(x)][1 + \xi_2'' K_2(x)] + \ldots + [1 + \xi_M'' K_M(x)] \\ & \ldots \text{etc.} \end{aligned}$$

the primes distinguish between successive passes through the set of g's and kernels; each pass may be regarded as a cycle of iterations. Thus $f'(x)$ can be written as:

$$f'(x) = 1 + (\xi_1 + \xi_1' + \xi_1'' + \ldots)\, K_1(x) + (\xi_2 + \xi_2' + \xi_2'' \ldots)\, K_2(x) + \ldots$$

$+ (\xi_M + \xi_M' + \xi_M'' \ldots)\, K_M(x)$ + terms involving products of ξ's and products of kernels

Hence as the iteration proceeds a linear combination of kernels is gen-

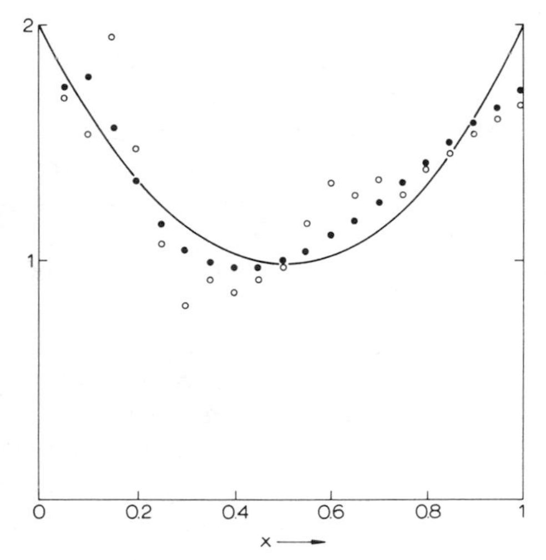

Fig. 7.7. Solution of the standard problem by Chahine method (open circles) and equation [7.21] (solid circles).

erated, plus various product combinations. The solution remains primarily within the function space spanned by the kernels, which in itself is beneficial for ensuring stability.

In Fig. 7.7. the solutions to the standard problem obtained by the Chahine procedure and by the method of equation [7.13a] are shown. They are similar in quality to those obtained by other methods, and as applied to this particular problem these more time consuming iterative methods possess no special advantages. The speed of improvement, as illustrated in Fig. 7.8, is

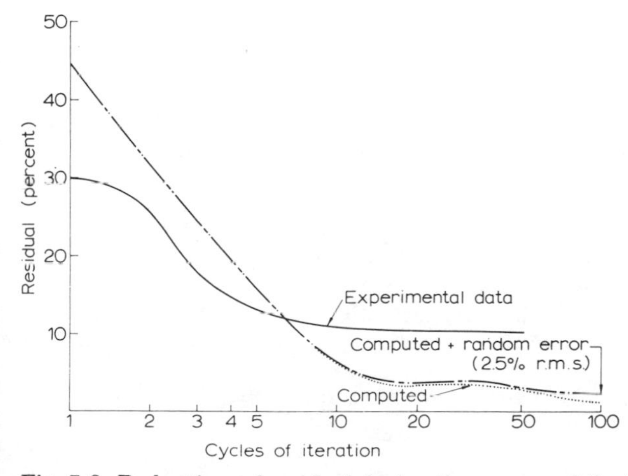

Fig. 7.8. Reduction of residual during the course of the iteration.

quite fast in the early stages but soon becomes very slow. In the author's experience, it does not seem likely that these methods are often likely to reduce residuals much below 1% (r.m.s.), but this is not a severe limitation since measurements are rarely better than this. It seems that the greatest advantage of these methods is their ability to cope with problems where the "dynamic range" of the unknown $f(x)$ is large — that is, where the solution ranges over a few orders of magnitude as x varies through its range. The more direct methods often fail in such situations, unless some transformation can be found to produce a modified $f(x)$ which does not vary so widely.

Before concluding this chapter it may be of some interest to compare the relative time taken to solve the standard problem by (1) constrained linear inversion with five values of γ — this is about how many γ's are usually needed at the exploratory stage; once a problem is established in detail, only a single γ may be needed; (2) Gilbert-Backus inversion with a similar number of values of γ; (3) iterative inversion by equation [7.13a], proceeding until the residual fell below 1%:

(1) constrained inversion ~ 1 time unit

(2) Gilbert-Backus ~ 60 time units

(3) non-linear iterative inversion ~ 4 time units

These times relate to the same problem in each case, but the programs were not refined to the limit in any instance. They are given purely as a guide to illustrate the relative speeds of the methods. From the point of view of programming time, the non-linear inversion procedure is very much simpler and quicker than the linear methods, of which the Gilbert-Backus method is somewhat more involved and therefore produces the longer program.

BIBLIOGRAPHY

Backus, G. and Gilbert, F., 1970. Uniqueness in the inversion of gross earth data. *Philos. Trans. R. Soc. Lond., Ser. A*, 266: 123—192.

Chahine, M.T., 1970. Inverse problems in radiative transfer: determination of atmospheric parameters. *J. Atmos. Sci.*, 27: 960—967.

King, J.I.F., 1964. Inversion by slabs of varying thickness. *J. Atmos. Sci.*, 21: 324—326.

Lanczos, C., 1956. *Applied Analysis*. Prentice-Hall, Englewood Cliffs, N.J., 539 pp.

Landweber, L., 1951. An iteration formula for Fredholm integral equations of the first kind. *Am. J. Math.*, 73: 615—624.

Strand, O.N. and Westwater, E.R., 1968. The statistical estimation of the numerical solution of a Fredholm integral equation of the first kind. *J. Assoc. Comput. Mach.*, 15: 100—114.

Whittaker, E.T. and Robinson, G., 1924. *The Calculus of Observations*. Blackie, London, 395 pp.

CHAPTER 8

Information content of indirect sensing measurements

An indirect sensing measurement may involve the measurement of (say) energy as a function of (say) wavelength — in other words, the energy in some set of narrow spectral intervals must be measured. Necessarily there is an error involved in this and almost always there is a degree of correlation among adjacent data points obtained by such measurement. (This, of course, must be the case if the very familiar process of drawing a smooth curve connecting experimentally or numerically obtained points is ever to be a justifiable procedure.) In this chapter these questions will be examined and some quantitative criteria will be derived whereby the question of correlation and interdependence of measured values can be examined objectively and related to the always-present errors.

8.1. HOW MANY MEASUREMENTS?

This question is very important from the practical point of view. If we contemplate any indirect sensing experiment, the inclination of most people is to add as many "independent" measurements as the budget or the engineering design (or perhaps in a satellite experiment weight or telemetry limitations) will allow. Forty, fifty or even more measurement points have been obtained in some measurements. It is obvious that no harm can be done by increasing the number of measurements if nothing else suffers as a result, but in many (if not most) real situations, there is a limiting and often quite small number of measurements, beyond which no *effective* increase in information is obtained.

This situation occurs when one of N measurements can be predicted mathematically from the others. Suppose the following numbers represent measurements of some y_i:

i	y_i
1	10.00
2	5.00
3	2.5
4	1.25
5	0.625
6	0.313
7	0.156
8	0.078
9	0.039
10	0.0195

There is evidently a high probability that an eleventh measurement would give 0.00975 ± 0.000005 since the y_i are obviously related to each other, y_i being to three significant figures just $\frac{1}{2}y_{i-1}$: a plot of y_i against i on log-linear graph paper would have given a straight line.

In such a situation an eleventh measurement would most probably be redundant to a high degree, unless by chance the relationship $y_i = \frac{1}{2}y_{i-1}$ happened to be an approximate one, which suddenly failed to hold for $i =$ 11. One cannot a priori and with complete certainty predict the outcome of the eleventh measurement without making it, but most people would regard the outcome as a foregone conclusion.

This sort of interdependence is of course empirical — we have written down a column of y's without giving them any physical meaning. But many of our physical measurement procedures do in fact imply some interdependence among the measurements; we think in terms of measuring some $f(t)$ at "close enough" intervals in t to "reproduce faithfully" the behavior of f as a function of t. If we measure $f = 1 \pm 0.000001$ at $t = 0.9999$, $t =$ 1.0000, $t = 1.0001$ there is still no absolute certainty that f does not go to infinity several times between 0.9999 and 1.0000. We in fact rely on intuition, or what we may know about the physics of $f(t)$ or of the measuring system to consider that the prescribed measurements "satisfactorily" measure $f(t)$. No number of measurements, however large, can give $f(t)$ in the same way that an analytic definition such as $f(t) = 10\ e^{-t} + 2\ e^{-3t}$, say, gives it. It is important to realize that there are two fundamental differences: (1) the measurements are finite in number, being made at a finite number of values of the argument, and (2) they are of finite accuracy.

Consider now indirect sensing measurements wherein measurements give a set of y_i, each y_i being an experimental value (subject to error) for $\int K_i(x) f(x)\, dx$, $f(x)$ being an inaccessible function and $K_i(x)$ a set of kernels which may also be subject to error (they may be experimental in origin or may be calculated by an approximate procedure). One can ask the question: given measurements of g_i (with some uncertainty due to error) for N values of i, *will additional information accrue from an additional measurement*? The answer, as one might expect, depends crucially on the nature of the kernels and on the measurement error. One might perhaps attempt to analyze the question by plotting g_i against y_i (y_i being the parameter, such as wavelength, that is varied to obtain different kernels) and observing how rapidly g_i changes with y_i. It would be reasonable to assume that if any three successively plotted points are rectilinear to within the accuracy of measurement, then there is no practical advantage to be gained from making a further measurement within that interval. Suppose the following data was obtained:

y_i	g_i
.	.
.	.
.	.
4.0	20.01
4.5	17.98
5.0	16.02
.	.
.	.

It is apparent that if these g_i are rounded to three figure accuracy, an exact linear relationship then exists between g_i and y_i. Hence there is no point whatever in adding further measurements between $y = 4$ and $y = 5$, unless the accuracy is such that we can rely on the fourth significant figure in the three values of g_i; indeed if the accuracy is less than that even the middle measurement for $y_i = 4.5$ could be deleted. But if our measurements are accurate to one part in 10^4 or better, there are significant deviations from linearity across the interval and an intermediate measurement can meaningfully be added.

One must, however, question the generalization involved in these arguments. It might have happened that the y_i versus g_i relationship was linear only by accident and that for other measurements (i.e. other $f(x)$'s) the linear relationship would not be found. In other words, the linearity could be caused by the particular $f(x)$, and if such were the case one could never delete the measurement at $y_i = 4.5$ on the grounds of its redundancy. However, the linearity can also be caused by the kernels; if $K(4.5, x)$ coincides accurately with $\frac{1}{2}K(4, x) + \frac{1}{2}K(5, x)$ throughout the interval over which the integral $\int K(y, x)\, f(x)\, dx$ is evaluated, then for *all* $f(x)$, $g(4.5)$ will be just $\frac{1}{2}g(4) + \frac{1}{2}g(5)$, and the redundancy of the middle measurement will *not* be accidental.

8.2. INTERDEPENDENCE OF KERNELS

We must therefore look at the possibility of writing any kernel as a linear combination of others. If for some y_i, say y_m, we can write:

$$K(y_m, x) = K_m(x) = \sum_{j \neq m} a_j K_j(x)$$

if this is exactly true for all relevant values of x, then of course:

$$g_m = \sum_{j \neq m} a_j g_j$$

and the mth measurement is totally redundant. In practice such exact relationships are unlikely, but error is always present and clearly if we write:

$$g_m = \sum_{j \neq m} a_j g_j + \delta_m$$

then if δ_m is absolutely less than the uncertainty with which we can obtain g_m by measurement, we have redundancy — g_m can be predicted to within this uncertainty from other measurements and obviously if we measure it we do not improve our knowledge concerning the unknown $f(x)$. One might use such a measurement to check the internal consistency of the overall system (and such a check is of undoubted value) but from the point of view of narrowing down the possible range of the unknown $f(x)$ it is ineffective. We come therefore quite naturally to the question: how small can we make $\Sigma_j a_j K(y_j, x)$, i.e. $\Sigma_j a_j K_j(x)$? Obviously by choosing all the a's to be zero the sum will vanish identically but this is not what we want, so some constraint must be applied to the a's which will not allow all of them to become zero simultaneously. Noting that if all the a's are doubled, so also is δ, since δ is equivalent to $\Sigma_j a_j \epsilon_j$, we observe that the absolute magnitude of the a's is irrelevant and we lose no generality by choosing the constraint $\Sigma_j a_j^2 = 1$, a unit-norm constraint which is also useful in placing bounds on δ, since the Schwarz inequality implies that $|\delta|^2 = |\Sigma_j a_j \epsilon_j|^2 \leqslant \Sigma_j a_j^2 \Sigma_j \epsilon_j^2$ so with the choice $\Sigma_j a_j^2 = 1$, the upper bound of $|\delta|^2$ is $\Sigma_j \epsilon_j^2$, which for independent, randomly distributed errors can be written $N|\epsilon|^2$ where $|\epsilon|$ is the r.m.s. error. (Since $|\delta|$ therefore increases as $\sqrt{N}$, there is always some improvement when N is increased, but this improvement will be present even if the additional measurements are simply repetitions of previous measurements. The $\sqrt{N}$-produced improvement is simply that caused by the increase in accuracy which is obtained by repeating a measurement several times or by increasing the integrating time of a measurement.) Choosing the constraint $\Sigma a_j^2 = 1$, one now must determine how small $|\Sigma a_j K_j(x)|$ can be made subject to this constraint. As a measure of the smallness of this quantity — a function of x — the square norm $\int [...]^2 \, dx$ is appropriate. The problem then becomes that of determining the conditions for $q = \int [\Sigma a_j K_i(x)]^2 \, dx$ to attain an extremum, subject to the constraint $\Sigma a_j^2 = \boldsymbol{a}^* \boldsymbol{a} = 1$. The measure q is a quadratic form in the unknown a's, and we can readily ascertain the associated matrix if we collect the $K_i(x)$ into a vector:

$$\boldsymbol{k}(x) = \begin{bmatrix} K_1(x) \\ K_2(x) \\ \cdot \\ \cdot \\ K_N(x) \end{bmatrix}$$

The scalar quantity $\Sigma_j a_j K_j(x)$ is simply: $\boldsymbol{a}^* \boldsymbol{k}(x)$ or $\boldsymbol{k}^*(x) \boldsymbol{a}$ and the square of this can be written $\boldsymbol{a}^* \boldsymbol{k}(x) \, \boldsymbol{k}^*(x) \boldsymbol{a}$. But matrix-vector products are associative so we can write this as the product of a row vector $\boldsymbol{a}^*$, a matrix $\boldsymbol{k}(x) \boldsymbol{k}^*(x)$, and a column vector $\boldsymbol{a}$, i.e.:

$$[a^*\boldsymbol{k}(x)]\,[\boldsymbol{k}^*(x)a] \equiv a^*[\boldsymbol{k}(x)\boldsymbol{k}^*(x)]\,a$$

scalar	scalar	row vector	matrix	vector
(1 × 1)	(1 × 1)	(1 × N)	(N × N)	(N × 1)

Only the matrix is a function of x, so the full integral $q = \int [\Sigma_j\, a_j\, K_j(x)]^2$ dx (a scalar) can be written $\boldsymbol{a}^*\mathbf{C}\boldsymbol{a}$. $\mathbf{C}$ is the "covariance matrix" $\|\int K_i(x)\, K_j(x)$ d$x\|$, a symmetric matrix obtained by taking all pairings of the kernels and integrating their products over the range of x with which the inversion problem is concerned. *Since* $\mathbf{C}$ *is symmetric its eigenvalues are all real and its eigenvectors mutually orthogonal.*

8.3. EXTREMA OF QUADRATIC FORMS

Because of its importance we will obtain the solution — i.e. the set of a_j which make q a minimum and the value which q takes there — in two different ways. The result has been somewhat anticipated earlier, but for completeness we will give both derivations in full at this point.

(1) This method relies on the existence of the eigenvectors and their completeness. Take our $\boldsymbol{a}$ as a linear combination of the eigenvectors $\boldsymbol{u}_1, \boldsymbol{u}_2$, etc., of $\mathbf{C}$, so that:

$$\boldsymbol{a} = \mathbf{U}\beta \qquad [8.1]$$

$\mathbf{U}$ of course being the matrix containing in its columns the normalized eigenvectors $\boldsymbol{u}_1, \boldsymbol{u}_2, ..., \boldsymbol{u}_N$. The constraint $\boldsymbol{a}^*\boldsymbol{a} = 1$ implies $\beta^*\beta = 1$ because:

$$\boldsymbol{a}^*\boldsymbol{a} = \beta^*\mathbf{U}^*\mathbf{U}\beta = \beta^*\beta = 1$$

the mutual orthonormality of the eigenvectors implying $\mathbf{U}^*\mathbf{U} = \mathbf{I}$. The value of q is evidently $q = \boldsymbol{a}^*\mathbf{C}\boldsymbol{a} = \beta^*\mathbf{U}^*\mathbf{C}\mathbf{U}\beta$ but $\mathbf{U}^*\mathbf{C}\mathbf{U} = \mathbf{\Lambda}$, hence:

$$q = \beta^*\mathbf{\Lambda}\,\beta = \sum_i \lambda_i\beta_i^2 \qquad [8.2]$$

The problem of minimizing q subject to the constraint $\Sigma a_i^2 = 1$ now becomes a trivial one — that of minimizing $\Sigma_i\, \lambda_i\beta_i^2$ subject to the constraint $\Sigma_i\, \beta_i^2 = 1$. If λ_1 is the smallest eigenvalue, the minimum of q is evidently given by choosing $\beta_1 = 1$, β_2 and all other β's being necessarily zero; the value attained by q at the minimum is obviously λ_1. $\beta_1 = 1$ is equivalent to $\boldsymbol{a} = \boldsymbol{u}_1$, the eigenvector associated with λ_1, so we have the result:

The smallest value attained by the positive definite quadratic form $q = \boldsymbol{a}^*\mathbf{C}\boldsymbol{a}$ *subject to the constraint* $\boldsymbol{a}^*\boldsymbol{a} = \Sigma a_i^2 = 1$ *is given by the smallest eigen-*

value λ_1 of **C**; *the choice for* ***a*** *which makes q take this value is the normalized eigenvector associated with* λ_1

(2) This method is more direct but less informative. The minimum of q when $a^*a = 1$ is given by the absolute minimum of $q + l\Sigma_i\, a_i^2$, l being a Lagrangian (undetermined) multiplier. Taking partial derivatives of $q + l\Sigma_i\, a_i^2$ with respect to a_1, a_2, and so on in succession, we obtain for the extrema the system of equations $(\partial/\partial a_k)[a^*\mathbf{C}a + la^*a] = 0$; noting that $(\partial/\partial a_k)a$ gives a vector e_k with zero elements in all but the kth position (where there is a one), we find:

$$e_k^*\mathbf{C}a + a^*\mathbf{C}e_k + le_k^*a + la^*e_k = 0$$

$$e_k^*[\mathbf{C}a + la] + [\mathbf{C}a + la]^*e_k = 0$$

These are equations involving *scalar* quantities; the second term is merely the transpose of the first and since transposition leaves a scalar unchanged, the two terms are always equal and must vanish separately. The necessary and sufficient condition for an extremum in q is therefore:

$$e_k^*[\mathbf{C}a + la] = 0 \qquad (k = 1, 2, \ldots N)$$

but any vector can be expressed as a sum of e_k's, so the equation:

$$v^*[\mathbf{C}a + la] = 0$$

holds for any v. Hence:

$$\mathbf{C}a + la = 0$$

but the eigenvector relationship for **C** is:

$$\mathbf{C}u - \lambda u = 0$$

So we can identify l with $-\lambda$ and must choose an eigenvector of **C** to make the quadratic form q an extremum. The value of q at an extremum is:

$$q_{\text{ext}} = u^*\mathbf{C}u = \lambda$$

Hence q is a minimum when λ is the smallest eigenvalue $\lambda_{\min}$ and u an associated eigenvector.

The very important result just obtained in two ways shows a fundamental property of eigenvectors, which in fact could be used as a definition in place of the matrix relationship $\mathbf{A}u = \lambda u$. It can be extended to a more general statement known as the minimax principle which can be stated as follows:

Given a positive definite quadratic form $a^*\mathbf{C}a$ *and m non-trivial independent equations of constraint, such as* $\Sigma_k\, \alpha_k a_k = 0$, $\Sigma_k\, \beta_k a_k = 0$, *etc., then as we vary the constraints but keep them independent the minimum value attained by the quadratic form q will vary; the maximum value attained by* *this minimum is the $(m + 1)$th eigenvalue of the matrix* **C** *(taking the eigenvalues in ascending order). Similarly the minimum value attained by the maximum of q as we vary the m constraints is the $(m + 1)$th eigenvalue in decreasing order.* The proof of these properties will be found in Courant and Hilbert's text (1953, pp. 31—33).

8.4. APPLICATION TO THE INTERDEPENDENCE PROBLEM FOR THE KERNELS

When a linear combination of functions vanishes identically in the interval of concern (i.e. when the square norm $\int [\Sigma a_i \; K_i(x)]^2 \; dx$ is zero) any one of the $K_i(x)$ can be expressed exactly as a weighted sum of the others; the only exception occurs if a certain a_i happens to be zero, for in that event the substitution:

$$K_i(x) = -a_i^{-1} \sum_{j \neq i} a_j K_j(x)$$

becomes indeterminate — however, all the a_i cannot be zero at the same time since $\Sigma a_i^2 = 1$ and at least one of the K_i's will be given by the above substitution. In the following discussion we will assume that the function $K_i(x)$ which we attempt to approximate is not associated with a vanishing value of a_i: it will be convenient to specify that $K_l(x)$ is the one being approximated and that a_l is numerically largest of the a_i, which therefore must numerically be at least $N^{-1/2}$, N being the total number of functions in the set.

An exactly vanishing combination $\Sigma_j \; a_j \; K_j(x)$ is a sufficient and necessary condition for some g, say g_l, to be exactly predictable from the other g's using:

$$g_l = -a_l^{-1} \sum_{j \neq l} a_j \int K_j(x) \, f(x) \, dx = -a_l^{-1} \sum_{j \neq l} a_j g_j$$

This will occur when there is a zero eigenvalue of the covariance matrix **C**. This will rarely occur, but often a *very small* eigenvalue will be found, which one might reasonably use to obtain an *approximating* expression for $K_l(x)$, and thereby a predicting formula for g_l, which will have only a *very small error*. We can easily determine the error in such a formula for $K_l(x)$. The approximation:

$$K_l(x) \cong \sum_{j \neq l} (-a_j/a_l) \; K_j(x) \qquad [8.3]$$

evidently involves the error:

$$\delta(x) = K_l(x) - \sum_{j \neq l} (-a_j/a_l) \; K_j(x)$$

So:

$$|\delta(x)| = |a_l^{-1}|\,|\sum_1^N a_j K_j(x)| = |a_l^{-1}|\,\lambda^{1/2}$$

from the derivation of the optimum $\boldsymbol{a}$. Since a_l is at least $N^{-1/2}$, the square norm of the error in $K_l(x)$ is $N^{+1/2}\lambda^{1/2}$ or less.

The error in the prediction of g_l is not quite as simple — obviously this involves the object function $f(x)$ also and we will have to specify something about $f(x)$ before any estimate of the error in g_l can be made. In just about every practical inversion problem a priori considerations limit the range of possible f's — the average magnitude of $f(x)$, may be fairly well defined, or the maximum value may be limited by the physical nature of $f(x)$.

Discussion of the error in $\boldsymbol{g}$ is much simpler if it is restricted to situations where the problem is scaled appropriately. Since any equation:

$$g_i + \epsilon_i = \int K_i(x)\, f(x)\, \mathrm{d}x$$

can be scaled by factors α and β:

$$\alpha g_i + \alpha\epsilon_i = \int (\alpha/\beta)\, K_i(x)\, \beta f(x)\, \mathrm{d}x$$

we can scale the magnitude of $\boldsymbol{g}$, K and f in any convenient way. If the measurement errors are relative (i.e. if each g_i is measured with the same expected percentage error) it is appropriate to scale the problem so that the individual g_i are of order unity. Then the ϵ_i can be identified with the relative errors and if the r.m.s. relative error is e, $|\epsilon|^2 = Ne^2$, while $|\boldsymbol{g}|^2 \cong N$. We can scale the right-hand side so that $|f(x)|^2$ is of order unity and of course the integration limit always can be scaled to $(0, 1)$.

To relate the prediction error in g_l to the measurement error component we have to compare the approximation:

$$g_l \cong \sum_{j \neq l} (-a_j/a_l)\, g_j$$

with the exact relationship:

$$g_l + \epsilon_l = \int K_l(x)\, f(x)\, \mathrm{d}x$$

letting $f(x)$ range through all admissible functions, while excluding as far as possible all impossible (on a priori grounds) functions $f(x)$. A given g_l may be found to be independent to a prescribed degree of the other g's when $f(x)$ is allowed to range through all bounded, continuous $f(x)$, but if the real range of $f(x)$ (i.e. function subspace to which $f(x)$ is practically limited) is more restrictive, all practically occurring g_l may be predicted better than measurement accuracy, even though a degree of independence would exist if $f(x)$ was less restricted. For this reason the tendency is for the independence of the g's to be *over*-estimated in practice.

The prediction of g_l may be written as g_l' where:

$$g_l' = \sum_{j \neq l} (-a_j/a_l)\, g_j$$

while

$$g_j = \int K_j(x)\, f(x)\, dx - \epsilon_j \qquad (j = 1, 2, \ldots N)$$

Inserting $K_l(x) = \Sigma_{j \neq l}(-a_j/a_l)\ K_j(x) + \delta(x)$ into the equation for g_l one obtains:

$$g_l = \sum_{j \neq l} (-a_j/a_l)\, K_j(x)\, f(x)\, dx + \int \delta(x)\, f(x)\, dx - \epsilon_l$$

$$= g_l' + \int \delta(x)\, f(x)\, dx - \sum_j \left(\frac{a_j}{a_l}\, \epsilon_j\right) \qquad [8.4]$$

This shows g_l to be made up of three components, which are (in order of their appearance on the right side of the last equation):

(1) A component $g_l' = \Sigma_{j \neq l}(-a_j/a_l)g_j$ *which is entirely predictable*, being a combination of the other g_j and coefficients which are calculated from the kernels only and do not involve the unknown $f(x)$ at all.

(2) A component $\int \delta(x)\, f(x)\, dx$ *which depends on the unknown* $f(x)$ and may happen to be zero for some $f(x)$ but large for another. This term is *not predictable* and it is in this term that new information about $f(x)$ is contained. One has essentially measured the projection of the function $f(x)$ in a new but not orthogonal direction in function space, and this term prescribes how much variation we can find along this direction.

(3) *The error term*, resulting from uncertainties in our experimental or empirical determination of the g's.

If the variation which can be produced by the second term is greater than that associated with the third (error) term, new information is yielded by measuring g_l. Otherwise we can calculate g_l' and thereby obtain an estimate for g_l which is closer than we can measure it: the independent unpredictable component in g_l is buried in the error "noise" and is obliterated by it.

The second term $\int \delta(x)\, f(x)\, dx$ can be assigned bounds only to the extent that $f(x)$ is bounded (but in practice it usually is). Schwarz's inequality then sets an upper limit to this term:

$$\left| \int \delta(x)\, f(x)\, dx \right|^2 \leqslant \left| \int \delta^2(x)\, dx \right| \left| \int f^2(x)\, dx \right|$$

but often this bound is very much larger than any value attained in practice. In many problems a more realistic (but less rigorous) estimate is obtained by using the mean value theorem to obtain:

$$|\int \delta(x)\, f(x)\, dx|^2 = f_m^2 |\int \delta(x)\, dx|^2 \leqslant f_m^2 \int [\delta(x)]^2\, dx$$

and inserting the approximate magnitude expected in $f(x)$ for f_m. The final step is to use a result earlier obtained which gave $\int [\delta(x)]^2\, dx = a_l^{-2}\lambda$. We can now write for the magnitude of the second term (recalling that $a_l \leqslant 1$):

$\leqslant |f|^2 \lambda$ (Schwarz's inequality)

or

$\leqslant |f_m|^2 \lambda$ (using mean value)

When the appropriate scaling has been done it is evident that the minimum eigenvalue represents an upper bound to the square norm of the unpredictable second component in equation [8.4].

The third term in equation [8.4] is that caused by errors in measurement or computation of the g_i and includes errors in the kernels themselves. Since an observation subject to errors may happen to coincide exactly with the true value, there is no minimum (other than zero) to the square magnitude of this term. However, it is clearly unrealistic to argue that all g_i are therefore independent of each other just because there is a small chance of measuring a set of g_i with no error. Neither is it realistic to derive an upper bound for the error term and discount any measurement which does not give an unpredictable component in g_l which is larger than the upper bound. The most useful quantity is some estimate of the *likely* magnitude of the error component $\Sigma_j\, a_l^{-1}\, a_j\epsilon_j$. The upper bound of this component is given by the Schwarz inequality as:

$$|a_l^{-1} \sum_j a_j\epsilon_j|^2 \leqslant a_l^{-2} \sum \epsilon_j^2 \qquad (\sum a_j^2 \text{ being unity})$$

but this is not necessarily a realistic estimate of the probable magnitude of that component. However, if there is no correlation between the error vector $\boldsymbol{\epsilon}$ and the unit vector $\boldsymbol{a}$, the expected norm of $\Sigma a_j\epsilon_j$, which is the length of the projection of the random vector $\boldsymbol{\epsilon}$ on the unit vector $\boldsymbol{a}$, is evidently $N^{-1}|\epsilon|^2$ which is the mean square relative error e^2. The value of a_l must lie between $N^{-1/2}$ and 1, and except for very large N is not too far from unity. Thus the error component $a_l^{-1}\, \Sigma_j\, a_j\epsilon_j$ has a magnitude not too different from the error magnitude $|\epsilon|$. Thus:

Provided the system is properly scaled, the independence of N measurements in the presence of a relative error of measurement $|\epsilon|$ is assured if $\lambda_{\min} > N^{-1}|\epsilon|^2$.

This is an especially important result. It can be extended to a more general statement if we note that the mutual orthogonality of the eigenvectors of $\|\int K_i(x) \int K_j(x)\, dx\|$ has the consequence that if m eigenvalues are less than $|\epsilon|^2$, then there are m redundant measurements which can be predicted as well as they can be measured.

If, for example, there are two eigenvalues less than $|\epsilon|^2$, then writing $\boldsymbol{a}$ and $\boldsymbol{b}$ for the corresponding eigenvectors, we have:

$$a_1 g_1 + a_2 g_2 + \ldots + a_N g_N = \delta_1 \cong 0$$

$$b_1 g_1 + b_2 g_2 + \ldots + b_N g_N = \delta_2 \cong 0$$

Elimination of (say) g_2 between these two equations gives a relationship expressing g_1 in terms of $g_3, g_4, \ldots, g_N$, and elimination of g_1 gives a relationship giving g_2 in terms of $g_3, g_4, \ldots, g_N$. (Note that all but one a_i can be zero, and all but one b_i can also be zero, but the two only non-vanishing elements cannot have the same value of i, since then $\boldsymbol{a} \cdot \boldsymbol{b}$ could not be zero. The orthogonality of the eigenvectors in fact assures the non-singularity of the elimination process.)

The presence of inequalities in the above results may give an impression of imprecision, but the reader is reminded that any discussion of errors cannot be precise in the usual sense. However, the conclusions which can be drawn from applying the results are almost always quite clear-cut. We must bear in mind that inequalities such as $\lambda > e^2$ should for practical purposes be written:

$$\lambda >> e^2$$

since if λ and e^2 are of the same order of magnitude, then there is a considerable amount of uncertainty about the corresponding inference, which is obtained with a "signal-to-noise ratio" of the order of unity. If the result obtained has an element of the unexpected — if, for example, the inferred $f(x)$ has a structure which one would not expect on the basis of its physical nature — then it could only be regarded as interesting but highly speculative until it is confirmed or supported in some independent way or the indirect measurement be improved sufficiently to remove the uncertainty.

Examples of the application of the interdependence tests

In Table 8.1 are set out the eigenvalues corresponding to the kernels shown in Fig. 8.1. These kernels although smooth, are well spaced along the x-axis, their maxima are separated and each kernel "reveals" a new strip of the x-axis: their appearance suggests a considerable degree of independence, and this impression is confirmed by the eigenvalues, as the table shows. The kernels are normalized to unity so that $\int_0^1 K_i^2(x)\, dx = 1$ for all i.

Provided the relative accuracy of measurement is a little better than 1%, the table shows that measurements made of $\int K_i(x) f(x)\, dx$ with $i = 1, 2, \ldots, 7$

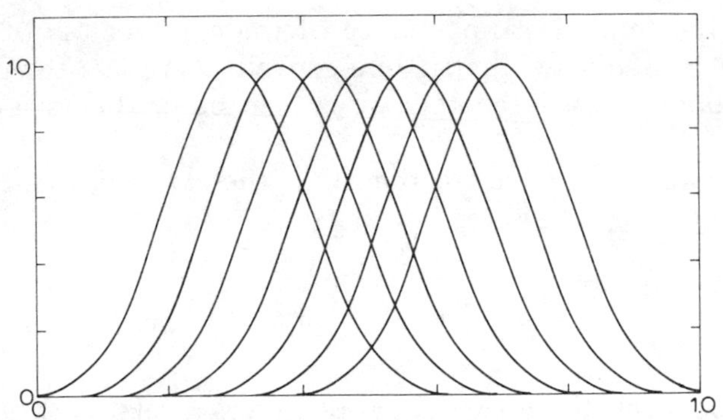

Fig. 8.1. Kernels consisting of equally spaced Gaussian functions. Information content for these kernels is high because each kernel uncovers a strip of the x-domain. Analysis of independence given in Table 8.1.

TABLE 8.1

Eigenvalues of the kernels shown in Fig. 8.1

i	λ_i	$\sqrt{\lambda_i}$
1	4.1749	2.04326
2	2.6375	1.6240
3	0.6375	0.7984
4	0.1320	0.3633
5	0.01822	0.1350
6	0.0001585	0.03982
7	0.00006827	0.008262

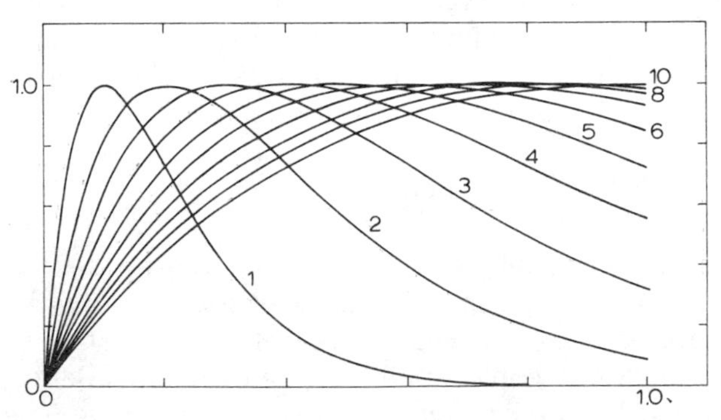

Fig. 8.2. Kernels for the "standard problem". These are similar to Poisson curves, and are more typical of most physical inversion kernels. The maximum moves along the x-axis but the curves become gentler and do not uncover new parts of the x-domain. Analysis given in Table 8.2.

are independent and each such measurement conveys new information about $f(x)$.

If, however, we take the kernels shown in Fig. 8.2 — exponential kernels e^{-yx} with $y^{-1} = 0.1, 0.2, \ldots, 1.0$ — we find a very different behavior of the eigenvalues (Table 8.2). Thus ten measurements of $\int_0^1 K_i(x)\, f(x)\, dx$ would have to be made to quite impossible accuracy (better than 10^{-7}) to make them independent. With 1% accuracy, eight of the measurements would be totally redundant.

The extent to which eigenanalysis can be applied to predict redundant measured data or to synthesize the corresponding kernels can be illustrated in a number of ways. Table 8.3 shows a set of actual measured data. (These were radiance values for wavelength intervals used in an early prototype of a satellite temperature profile sensing instrument.) The kernels were typical smooth atmospheric transmission kernels (Fig. 8.3). The eigenanalysis of these kernels yielded the following data:

Eigenvalues:

1.84×10^{-4}, 3.84×10^{-3}, 7.33×10^{-2}, 0.595, 7.18, 634

Eigenvectors:

$$\begin{bmatrix} 0.00365 \\ -0.0978 \\ 0.381 \\ -0.654 \\ 0.602 \\ -0.236 \end{bmatrix} \quad \begin{bmatrix} -0.0338 \\ 0.310 \\ -0.647 \\ 0.212 \\ 0.540 \\ -0.384 \end{bmatrix} \quad \begin{bmatrix} -0.667 \\ 0.382 \\ 0.445 \\ 0.249 \\ -0.093 \\ -0.376 \end{bmatrix} \quad \begin{bmatrix} 0.624 \\ 0.418 \\ 0.132 \\ -0.128 \\ -0.362 \\ -0.520 \end{bmatrix} \quad \begin{bmatrix} 0.350 \\ 0.395 \\ 0.410 \\ 0.421 \\ 0.431 \\ 0.437 \end{bmatrix}$$

The first-listed eigenvector, that associated with the smallest eigenvalue, has its numerically largest element in the fourth position. The best approx-

TABLE 8.2

Eigenvalues of the exponential kernels of Fig. 8.2

i	λ_i	$\sqrt{\lambda_i}$
1	6.28	2.506
2	0.04	0.1988
3	4.2×10^{-5}	0.000648
4	1.8×10^{-8}	0.00013
5	3.7×10^{-12}	1.92×10^{-6}
6	$<10^{-14}$	$<10^{-7}$
7	$<10^{-14}$	$<10^{-7}$
8	$<10^{-14}$	$<10^{-7}$
9	$<10^{-14}$	$<10^{-7}$
10	$<10^{-14}$	$<10^{-7}$

TABLE 8.3

Comparison of the exact values of K_4 with its approximation by means of a linear combination of the other kernels

p (mbar)	K_4 — exact	Approximation	p (mbar)	K_4 — exact	Approximation
0.100	1.000	1.000	64.8	0.645	0.646
1.13	0.991	0.989	78.4	0.583	0.582
2.31	0.982	0.981	90.4	0.531	0.529
3.73	0.972	0.971	109	0.456	0.453
5.72	0.959	0.957	139	0.354	0.350
10.1	0.932	0.932	168	0.269	0.268
14.1	0.909	0.908	194	0.208	0.212
20.7	0.871	0.870	234	0.133	0.146
27.5	0.833	0.834	283	0.070	0.079
36.6	0.784	0.786	360	0.020	0.014
44.3	0.744	0.745	415	0.007	−0.011
53.6	0.698	0.699	552	0.000	−0.025

imation is therefore that obtained by setting the combination equal to zero and solving for $K_4(x)$. This gives:

$$0.654K_4(x) \cong 0.00365K_1(x) - 0.0978K_2(x) + 0.381K_3(x) + 0.602K_5(x) - 0.236K_6(x)$$

i.e.:

$$K_4(x) \cong 0.00558K_1(x) - 0.1496K_2(x) + 0.583K_3(x) + 0.921K_5(x) - 0.360K_6(x)$$

and hence for any measured g:

$$g_4 \cong 0.00558g_1 - 0.1496g_2 + 0.583g_3 + 0.921g_5 - 0.360g_6$$

Table 8.3 compares the values of the above approximation for $K_4(x)$ with its exact values. (Physically x was the logarithm of the atmospheric pressure p, which is similar to a height variable.) Note that most entries for $K_4(x)$ differ only in the third significant figure from the predicted value given by the appropriate linear combination of $K_1(x)$, $K_2(x)$, $K_3(x)$, $K_5(x)$ and $K_6(x)$.

The eigenvectors given above deserve some attention. The largest eigenvalue is associated with an eigenvector with elements which are all roughly equal. Thus what is "seen" most easily is the dot product of the atmospheric profile with the average of the kernels. The smallest eigenvalue, on the other hand, is associated with an eigenvector which has elements alternating in sign and the associated orthogonal function $\phi_5(x) = \Sigma_n\, u_{n5}\, K_n(x)$ will be the most oscillatory; this "sees" the finest structure in $f(x)$ but the smallness of

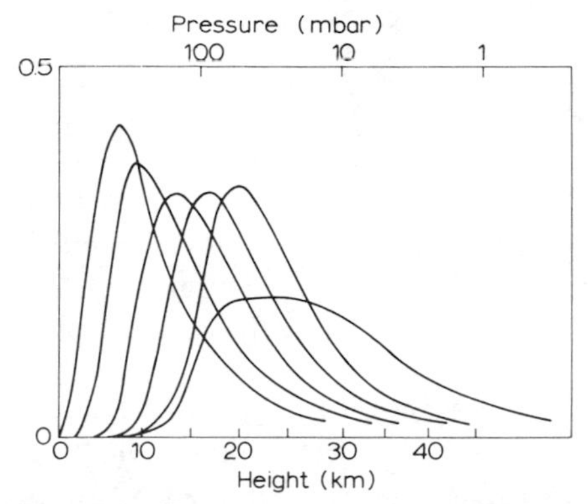

Fig. 8.3. Kernels for an actual atmospheric temperature sensing experiment using six wavelength intervals in the 15-μm CO_2 band. The quantity plotted is the effective kernel — the derivative of the transmittance (see Chapter 1).

the eigenvalue implies that this is "seen" only very weakly and can only be extracted by very accurate measurements.

8.5. INFORMATION CONTENT

In the preceding pages an objective method was developed for analyzing a set of kernels (measured, computed or analytic, so long as the covariance matrix $\|\int K_i(x)\, K_j(x)\, dx\|$ can be calculated) to ascertain the degree of interdependence existing among them. When M measurements of a total of N measurements are not predictable to within the measurement error, one can speak of M pieces of independent information existing among the N measurements — the remaining $N-M$ "theoretical" pieces of information are lost in the error noise and not amenable to mathematical retrieval.

When N is increased, M may or may not increase also. As an example the "standard" kernels will be used: functions of the form $x\, e^{-yx}$, with y ranging from one to infinity, so that the maximum of the function (which occurs at $x = y^{-1}$) ranges between 0 and 1. For each value of N, N kernels are constructed so as to give maxima that are uniformly spaced between $x = 0$ and $x = 1$; for $N = 4$, for example, the kernels chosen were:

$K(x) = x\, e^{-x}$ (maximum at $x = 1$)

$K(x) = x\, e^{-4x/3}$ (maximum at $x = \frac{3}{4}$)

$K(x) = x\, e^{-2x}$ (maximum at $x = \frac{1}{2}$)

$K(x) = x\, e^{-4x}$ (maximum at $x = \frac{1}{4}$) .

The kernels were first normalized to unit area, using the relationship:

$$\int_0^1 x\, e^{-yx}\, dx = y^{-2}(1 - e^{-y} - y\, e^{-y})$$

and then were used to obtain an N-order covariance matrix $\mathbf{C} = \|\int_0^1 K_i(x)\, K_j(x)\, dx\|$ which was subjected to an eigenanalysis using the Jacobi orthogonalization procedure. (The results were checked by calculating $\mathbf{UU}^*$ and comparing this with the original matrix $\mathbf{C}$, and were found to give agreement to better than six significant figures.)

The results are set out schematically in Fig. 8.4 where the eigenvalues are indicated on a logarithmic scale for various values of N. For smaller values of N all of the eigenvalues have been plotted but for larger values of N some of the eigenvalues were so small that their inclusion would serve no useful purpose. When the number of dots in the figure does not equal the value of N, the missing eigenvalues were smaller than the smallest value accommodated on the graph. The figure shows that merely increasing N does not achieve much increase in information content — a considerable increase in measurement accuracy is called for, something easier said than done in most situations. The kernels used in this example were hypothetical for convenience of

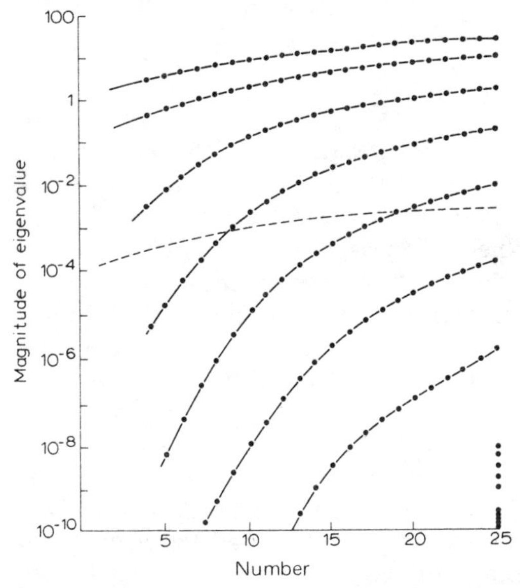

Fig. 8.4. The effect of increasing the number of measurements by increasing the number of y_i- values in the fixed interval. The number of useful pieces of information is given by the number of eigenvalues which exceed a fixed value that depends on the accuracy with which the measurements have been done. The broken line represents this limit for 1% accuracy. Upper curve gives largest eigenvalue, next curve down the next largest, etc. Number of points lying above the broken line gives the information content for 1% accuracy.

computation, but the behavior seen in Fig. 8.4 is very typical of practical physical kernels. For a set of kernels related to an optical light-scattering experiment wherein particle size distribution was to be inferred from back-scattering measurements, results very similar to those of the figure were obtained (Table 8.4).

It is important to emphasize what the results of an information content analysis imply. If we conclude that five pieces of information are available from a set of measurements with a prescribed accuracy, it does *not* mean that *any* five pieces of information can be obtained. One may categorize a function $f(x)$ by its value for various values of x, by its Fourier coefficients, by its moments, and so on, but the existence of five pieces of information does not mean that $f(x)$ for some fixed five values of x, or the first five moments or five Fourier coefficients of $f(x)$ are necessarily obtainable.

This point is seen most readily from the viewpoint of function space. The measurements of $\int K_i(x)\, f(x)\, dx$ provide values of the projection of the unknown $f(x)$ on the skewed system of axes $K_1(x)$, $K_2(x)$, etc. An orthogonal system of axes is provided by the functions $\phi_1(x)$, $\phi_2(x)$, etc., formed by combining the functions $K_i(x)$ with elements of eigenvectors of the covariance matrix as coefficients, i.e.:

$$\phi_1(x) = u_{11}K_1(x) + u_{21}K_2(x) + \ldots + u_{N1}K_N(x)$$

$$\phi_2(x) = u_{12}K_1(x) + u_{22}K_2(x) + \ldots + u_{N2}K_N(x)$$

... etc.

These functions, however, are not normalized — their square norms are the respective eigenvalues and to obtain an orthonormal set we must use the functions $\hat{\phi}_k(x) = \lambda_k^{-1/2}\, \phi_k(x)$. To obtain the projections of $f(x)$ on this orthonormal system the measurements must be multiplied by numbers $\lambda_k^{-1/2}\, u_{lk}$, which may be very large when the eigenvalues are very small. For small enough eigenvalues the true value of the projection becomes dominated by the error component (which is magnified by the $\lambda_k^{-1/2}$ factor) and is therefore inaccessible. The function subspace spanned by the correspond-

TABLE 8.4

Information content analysis in a light-scattering problem

N	Information content (pieces of information)		
	for ~3% r.m.s. error	for ~1% r.m.s. error	for 0.3% r.m.s. error
5	5	5	5
10	5	6	6
15	5	6	7
22	5	6	7

TABLE 8.5

Accuracy of synthesis of the weighting functions 1, x, x^2, ... by means of linear combinations of 22 optical transmission kernels

Weighting function $w(x)$	Producing	Relative error in $w(x)$ with 22 exact measurements (%)
1	population number	28
x	mean	8
x^2	variance	1.4
x^3	skewness	0.9
x^4	kurtosis	0.7

ing ϕ's is inaccessible, and any data involving this part of the subspace is not available. In general the functions $\phi_k(x)$ will not coincide with any of the functions 1, x, x^2, ... which are base functions for the moments of $f(x)$ (i.e. the moments are the inner products of $f(x)$ with those base functions) or the functions 1, $\cos x$, $\sin x$, $\cos 2x$, $\sin 2x$, etc., which generate the Fourier coefficients; the $\phi_k(x)$ never can coincide with the delta function $\delta(x-u)$ which gives the value of $f(x)$ at $x = u$.

For these reasons we cannot always obtain desired information about $f(x)$ even though a number of independent pieces of information are available: in fact, a desired $\int \psi(x)\, f(x)\, dx$ will only be accessible to the extent that $\psi(x)$ lies within the accessible part of the function subspaces defined by $\phi_k(x)$'s (where the information is not completely obscured by magnified error noise). This implies that if there are M pieces of information contained in a given set of measurements and $\phi_1(x)$, $\phi_2(x)$, ..., $\phi_M(x)$ are the corresponding orthogonal base functions then we can usefully determine $\int \psi(x)$ $f(x)\, dx$ provided:

$$|\psi(x) - [\psi \cdot \phi_1]\, \phi_1(x) - [\psi \cdot \phi_2]\, \phi_2(x) \ldots - [\psi \cdot \phi_m]\, \phi_m(x)|$$

is small compared to $|\psi(x)|$, that is, *too much of* $\psi(x)$ *must not lie outside the accessible function subspace.*

In many instances the kernels are such that all the lower moments of $f(x)$ can be inferred with confidence, but this is not always true. In optical transmission kernels, for example, the lower moments are not all accessible, for the measurements are quite insensitive to the moments $\int f(x)\, dx$ and $\int x f(x)\, dx$, as shown by the data in Table 8.5, relating to optical scattering methods for indirect measurement of particle size distribution. There is here evidently a deficiency in the very lowest moments.

8.6. INDEPENDENCE ANALYSIS APPLIED TO MEASURED QUANTITIES

Up to now we have concentrated on the degree of independence existing among the kernels and the extent to which one kernel can be expressed as a combination of the others. But one can also apply the analysis directly to *measurements*. If a certain $g(y)$ is measured at a large number of values of y, $g(y)$ still may only contain a small number of *independent* pieces of information. If $g(y)$ is related to $f(x)$ by the indirect sensing equation $g(y) = \int K(y, x)\, f(x)\, dx$ and only a few of the kernels are needed to provide an accurate approximation of the kernels for all other relevant values of y, then obviously the same situation will exist as far as $g(y)$ is concerned — there will be a strong correlation among the g's and $g(y)$ for any relevant y will be closely approximated by a linear combination of a small number of values: given some suitably located $g(y_1)$, $g(y_2)$, ..., $g(y_n)$ we have in essence an interpolating or extrapolating formula which gives us values of $g(y)$ for all other values of y in the interval. The effectiveness of the prediction obtained by such formulae has already been shown, but the formula was derived from the kernels. Had a large number of sets of measurements of $g(y)$ at closely spaced values of y been available, it would have been possible to derive the formula and to examine the degree of independence among the g's without any reference to the kernels.

The point at issue here is that if a large set of $g(y)$'s are (to within measurement accuracy) merely combinations of a few basic functions, then the information content of the g's is set by the number of independent basic functions which are blended together to form the g's — not by the number of tabular values at which the $g(y)$'s are measured. If, for example, a certain function happened to be $y^3 + 2\cos(y - 1) + 3\, J_0(\pi y)$, a plot of this function would exhibit a considerable amount of structure and it would be almost impossible to recognize it as a combination of three basic patterns, but even if it was calculated or measured for a thousand values of y, this does not represent a thousand pieces of independent information.

To determine directly how many pieces of information are contained in a set of functions $g_i(y)$, we note that if that set represents a linear combination of M basic patterns, then the basic patterns $p_1(y)$, $p_2(y)$, etc., will be linear combinations of the $g_i(y)$, so that:

$$\left.\begin{aligned} p_1(y) &= b_{11}g_1(y) + b_{12}g_2(y) + \ldots + b_{1N}g_N(y) \\ p_2(y) &= b_{21}g_1(y) + b_{22}g_2(y) + \ldots + b_{2N}g_N(y) \\ &\text{etc.} \end{aligned}\right\} \qquad [8.5]$$

We can assume without loss of generality that the $p_i(y)$ are orthogonal and normalized, for if a given $p_k(y)$ is not orthogonal to each of the preceding p's, we can remove the non-orthogonal component from $p_k(y)$ and incorporate it into terms containing lower order p's; the remaining part of $p_k(y)$, if not zero, is orthogonal to the preceding p's — if it is zero $p_k(y)$ is redundant and can be deleted. Orthonormality of the p's implies that $\int p_m(y)\, p_n(y)\, dy = 0$ $(m \neq n)$ and $= 1$ $(m = n)$, or equivalently that the matrix

$\int p(y)\, p^*(y)\, dy$ is the identity matrix; equation [8.5] can be written:

$$p(y) = \mathbf{B}g(y)$$

here $p(y)$ may be taken as a vector of dimension N, as is $g(y)$, but we have no guarantee that N distinct orthonormal $p_i(y)$ can be obtained from an arbitrary set of $g(y)$; it may be necessary to repeat some of the equations to obtain N equations in equation [8.5] and of course in that event the matrix $\mathbf{B}$ will be singular, since its determinant, having two or more equal rows, must vanish.

The orthonormality condition $\int p(y)\, p^*(y)\, dy = \mathbf{I}$ implies $\mathbf{B}\int g(y)\, g^*(y) dy\, \mathbf{B}^* = \mathbf{BCB}^* = \mathbf{I}$ (the use of $\mathbf{C}$ is appropriate since $\int g(y)\, g^*(y)\, dy$ is the covariance matrix of the $g_i(y)$). Hence $\mathbf{B}$ is just $\Lambda^{-1/2}\, \mathbf{U}^*$, $\mathbf{U}$ being as usual the matrix which contains in its columns the eigenvectors of $\mathbf{C}$: $\mathbf{B}$ therefore contains the eigenvectors u_1, u_2, etc., multiplied by $\sqrt{\lambda_1}$, $\sqrt{\lambda_2}$, etc., in its rows. Hence if $\mathbf{C}$ is non-singular we can obtain N independent orthonormal patterns, i.e. the $g_i(y)$ are independent on the relevant interval. An equivalent statement is that the independence of $g_1(y), g_2(y), ..., g_N(y)$ is assured if the determinant:

$$G = \begin{vmatrix} \int g_1^2 dy & \int g_1(y)\, g_2(y)\, dy & \int g_1(y)\, g_3(y)\, dy & \cdots \\ \int g_1(y)\, g_2(y)\, dy & \int g_2^2(y)\, dy & \int g_2(y)\, g_3(y)\, dy & \cdots \\ \cdot & \cdot & \cdot & \\ \cdot & \cdot & \cdot & \end{vmatrix} \qquad [8.6]$$

does not vanish. This determinant is known as the "Gram determinant" for the set of functions $g_1(y), g_2(y), ..., g_N(y)$.

The above criterion is an analytic, exact one. The Gram determinant will vanish only if one or more linear combinations of the g's vanishes identically. But for practical purposes we need a stronger criterion: any combination which is small compared to the error noise in the g's should be discarded. In other words, if we can form a combination $\Sigma_j\, b_j g_j(y)$ which is less in magnitude than the expected magnitude of the corresponding error $\Sigma_j\, b_j \epsilon_j(y)$, this is no better than an exactly vanishing combination. Such combinations will occur when:

$$\int \{\sum_j b_j g_j(y)\}^2\, dy < \overline{\int \{\sum_j b_j \epsilon_j(y)\}^2\, dy}$$

or:

$$b^* \mathbf{C} b < b^* \| \overline{\int \epsilon_i(y)\, \epsilon_j(y)\, dy} \|\, b \qquad [8.7]$$

the bars indicate averaging. If the errors are random with equal expectation all but the diagonal elements in the last matrix vanish after averaging and the diagonal elements are all equal to the average square norm of the error, $|\epsilon|^2$, so the last matrix can then be written as $|\epsilon|^2\mathbf{I}$ and the inequality becomes (putting for b an eigenvector of $\mathbf{C}$):

$$\lambda < |\epsilon|^2$$

For every eigenvalue of the $N \times N$ matrix $\mathbf{C}$ which is less than $|\epsilon|^2$ the information content must be decreased by one. The number of non-redundant independent patterns contained in the set of N functions $g_1(y)$, $g_2(y)$, ..., $g_N(y)$ is simply the number of eigenvalues of the covariance matrix which exceed $|\epsilon|^2$, the average square norm of the errors in the $g_i(y)$.

An interesting and instructive application of this result was that made by Mateer (1965), who examined records obtained for many years by the Umkehr procedure for indirectly estimating ozone profiles in the atmosphere. Umkehr measurements had been made in many parts of the world, sometimes in very great detail (fifty or more measurements as the solar angle varied), but Mateer showed that only some *four* basic patterns were involved. For a set of 100 Umkehr curves for Arosa, Switzerland, the analysis yielded the following eigenvalue data:

i	λ_i
1	0.978
2	0.01505
3	0.00529
4	0.00114
5	0.00034

When one examines a large number of functions by expanding them in terms of some orthogonal base functions, the overall variability contained in any one function can be measured by $\int [f(y)]^2\, dy$: this quantity corresponds to "power" in engineering or spectral analysis. Corresponding to a power spectrum, which describes how different frequencies contribute to the power, we can plot or tabulate the contribution of different orthogonal components to the value of $\int [g(y)]^2\, dy$. Since the orthonormal functions have vanishing cross-product terms we have:

$$\int [g_1(y)]^2\, dy = \int [\xi_{11}p_1(y) + \xi_{12}p_2(y) + \ldots]^2\, dy$$

$$= \xi_{11}^2 + \xi_{12}^2 + \ldots + \xi_{1N}^2$$

similarly, for all other g's:

$$\int [g_2(y)]^2 \, dy = \xi_{21}^2 + \xi_{22}^2 + \ldots + \xi_{2N}^2$$

.

.

$$\int [g_m(y)]^2 \, dy = \xi_{M1}^2 + \xi_{M2}^2 + \ldots + \xi_{MN}^2$$

Thus $\Sigma_{k=1}^{m} \int [g_k(y)]^2 \, dy$ represents the total "power" in the m functions; this equals $\Sigma_k \Sigma_l \xi_{kl}^2$ and the power contribution from the $p_1(y)$ components is $\Sigma_k \xi_{k1}^2$, the contribution from $p_2(y)$ components is $\Sigma_k \xi_{k2}^2$, and so on. Thus a plot of $\Sigma_k \xi_{kl}^2$ against l is a kind of generalized power spectrum which sets out the contributions of the various orthogonal patterns $p_1(x)$, $p_2(x)$, etc., to the "power" in the overall set of $g_i(y)$. Having derived the functions $p_1(x)$, $p_2(x)$, etc., it is a straightforward but laborious matter to calculate the coefficients ξ_{kl}, for these are given by $\int g_k(y) \, p_l(y) \, dy$, but as it happens this is unnecessary — the eigenvalues give the necessary information. We have:

$$p(y) = \Lambda^{-1/2} \mathbf{U}^* g(y)$$

So:

$$\int g(y) p^*(y) dy = \int g(y) g^*(y) \mathbf{U} \Lambda^{-1/2} dy$$

$$= \mathbf{C U} \Lambda^{-1/2} = \mathbf{U} \Lambda^{1/2}$$

Hence:

$$\xi_{kl} = u_{kl} \lambda_l^{1/2}$$

and

$$\sum_k \xi_{kl}^2 = \lambda_l \sum_k u_{kl}^2 = \lambda_l (u_l \cdot u_l) = \lambda_l$$

The lth eigenvalue of the covariance matrix $\mathbf{C} = \| \int g_i(y) \, g_j(y) \, dy \|$ gives the contribution of the lth orthogonal pattern $p_l(y)$ to the overall "power" $\Sigma_k \int [g_k(y)]^2 \, dy$.

Returning to Mateer's analysis, it now can be seen to imply that 97.8% of the overall variation in the Arosa Umkehr data is contributed by the first pattern, 1.505% by the second, 0.529% by the third and 0.114% by the fourth. In other words, 99.98% of the total variability has already been accounted for; the remaining 0.02% corresponds to a relative r.m.s. error component of about $\sqrt{0.0002}$ or a little over 1%, which is of the order of the experimental error.

The procedure described above is often called "principal component analysis". From the result just obtained it is clear that if the eigenvalues and vectors are arranged in decreasing order, there will be no better way of combining the original functions as judged from the viewpoint of how much of the overall $\Sigma_k \int [g_k(y)]^2$ dy is explained or accounted for at each step. That is, the expansion of the $g_i(y)$ in terms of the patterns $p_1(y)$, $p_2(y)$, etc., is *optimal*; when each expansion is truncated after a number of terms, a greater part of the quantity $\Sigma_k \int [g_k(y)]^2$ dy is already accounted for than with any other possible expansion of the $g_i(y)$. (We have not proved the optimality, but this property follows in a straightforward way from the extremal properties of eigenvectors.)

The variance of a set of m functions can be expressed as $\Sigma_k \int [g_k(y) - \overline{g(y)}]^2$ dy where $\overline{g(y)}$ is the mean $(1/m)\Sigma_k g_k(y)$. If the set of functions happens to have a zero mean, variance and "power" (as understood above) are synonymous. Otherwise we find a first pattern which is either the mean or a weighted mean, depending on how the original functions are normalized.

The results derived in this section can of course be applied to sets of functions $g_i(y)$ which are not related to indirect sensing or inversion problems. The procedures discussed are closely related to the statistical techniques of variance analysis and principal component analysis; the special orthogonal functions $p_i(x)$ which were found to be optimal for the expansion of the functions $g_i(y)$ are found under a variety of names of which the commoner are "principal components", "characteristic patterns" and "orthogonal empirical functions". In some instances it is possible to identify different patterns with different physical interactions or perturbations, but the primary character of the patterns is statistical — one must be careful about assuming that algebraic orthogonality will result from physically independent mechanisms; there is no assurance that statistical orthogonality guarantees anything else.

8.7. ERROR MAGNIFICATION

The results derived in [8.4] can be obtained in a slightly different way if a suitable inversion formula is derived and its application viewed in the light of error magnification. Writing the fundamental integral relationship, as usual, in the form:

$$g_i = \int_a^b K_i(x)\, f(x)\, dx \qquad (i = 1, 2, \dots M)$$

and seeking a solution for $f(x)$ in terms of some as yet unspecified orthogonal set of functions $\phi_1(x), \phi_2(x), \dots, \phi_n(x)$, we find:

$$g_i = \int_a^b K_i(x) \sum_j \xi_j \phi_j(x)\, dx$$

$$g = \int_a^b k(x)\,\phi^*(x)\,\xi \mathrm{d}x$$

if

$$f(x) = \sum \xi_j \phi_j(x) = \xi^* \phi(x) = \phi^*(x)\,\xi.$$

Since there is no possibility of "seeing" any component in $f(x)$ which is orthogonal to all of the kernels, the most general set of ϕ's is given by:

$$\phi(x) = \mathbf{W}^* k(x)$$

(as yet $\mathbf{W}^*$ is arbitrary, but the reason for writing it as a transpose will soon become obvious). The postulated orthonormality of the ϕ's implies that:

$$\mathrm{I} = \int \phi(x)\,\phi^*(x)\,\mathrm{d}x$$

$$= \mathbf{W}^* \int k(x)\,k^*(x)\,\mathrm{d}x\mathbf{W} = \mathbf{W}^*\mathbf{C}\mathbf{W}$$

($\mathbf{W}$ is not a function of x). The relationship $\mathbf{C} = \mathbf{U}\Lambda\mathbf{U}^*$ or $\mathbf{U}^*\mathbf{C}\mathbf{U} = \Lambda$ immediately provides us with the required $\mathbf{W}$, which obviously must be $\mathrm{U}\,\Lambda^{-1/2}$; this choice gives:

$$\mathbf{W}^*\mathbf{C}\mathbf{W} = \Lambda^{-1/2}\mathbf{U}^*\mathbf{C}\mathbf{U}\,\Lambda^{-1/2} = \Lambda^{-1/2}\,\Lambda\Lambda^{-1/2} = \mathrm{I}$$

as is required.

With these substitutions the fundamental integral relationship is transformed into:

$$g = \int_a^b k(x) k^*(x) \mathrm{d}x \cdot \mathrm{U}\,\Lambda^{-1/2}\,\xi$$

$$= \mathbf{C}\mathbf{U}\,\Lambda^{-1/2}\,\xi$$

or:

$$g = \mathrm{U}\,\Lambda^{1/2}\,\xi \qquad [8.9]$$

This can immediately be inverted, for $\mathbf{U}$, being the matrix of eigenvectors of a real symmetric matrix, is orthogonal and so:

$$\xi = \Lambda^{-1/2}\mathbf{U}^* g \qquad [8.10]$$

and:

$$f(x) = k^*(x)\,\mathrm{U}\,\Lambda^{-1}\mathbf{U}^* g \qquad [8.11]$$

To verify that $f(x)$ thus calculated satisfied the original integral relationship, we need only substitute $k^*(x)\ \mathrm{U}\Lambda^{-1}\ \mathbf{U}^* g$ for $f(x)$ into $\int k(x)\ f(x)\ \mathrm{d}x$; this gives:

$$\int k(x)\, f(x)\, dx = \int k(x) k^*(x) dx\, U \Lambda^{-1} U^* g$$

$$= C U \Lambda^{-1} U^* g = U U^* g = g$$

So [8.11] indeed gives a solution to the original equation.

This solution could be used as a method of inversion, but being exact it will of course be unstable. Its main value is that it allows one to look at the error magnification produced in a very straightforward way. When we add to g an error vector ϵ, this clearly produces in $f(x)$ an error:

$$\delta(x) = k^*(x)\; U \Lambda^{-1} U^* \epsilon$$

If the errors are random with equal variances and with square norm $\Sigma \epsilon_i^2 = \epsilon^* \epsilon = e^2$, then $U^* \epsilon$ is simply a "rotated" vector of random errors ϵ' with the same norm — for:

$$|\epsilon'|^2 = |U^* \epsilon|^2 = (U^* \epsilon)^*\; U^* \epsilon = \epsilon^* U U^* \epsilon = \epsilon^* \epsilon = |\epsilon|^2$$

— and the same probability. The square norm of the error $\delta(x)$ can be readily written down; it is:

$$|\delta(x)|^2 = \int_a^b [\delta(x)]^2\, dx = (\epsilon')^*\, \Lambda^{-1} U^* \int k(x) k^*(x) dx \cdot U \Lambda^{-1}\, \epsilon'$$

$$= (\epsilon')^* \Lambda^{-1} U^* C U \Lambda^{-1} \epsilon' = (\epsilon')^* \Lambda^{-1} \Lambda \Lambda^{-1} \epsilon'$$

So:

$$|\delta(x)|^2 = (\epsilon')^*\; \Lambda^{-1}\; \epsilon' \qquad [8.12]$$

This relationship once again emphasizes the crucial role of small eigenvalues in producing large errors in the solution from small errors in g. It also shows that, as one might expect, all combinations of errors or for that matter the same error in different places, do not have the same effect. If we write [8.12] in scalar form, it becomes:

$$|\delta(x)|^2 = \sum_i \lambda_i^{-1} (\epsilon_i')^2 = \lambda_1^{-1} (\epsilon_1')^2 + \lambda_2^{-1} (\epsilon_2')^2 + \dots \qquad [8.13]$$

so, if $\Sigma_i (\epsilon_i')^2$ happens to be unity, $|\delta(x)|^2$ will range between λ_1^{-1}, and λ_M^{-1}. This can represent an enormous range in error magnification — in some of the examples given earlier the eigenvalues of **C** ranged over ten decades or more, so one combination of error could produce a value of $|\delta(x)|^2$ which was 10^{10} or more times that produced by another, apparently comparable, set of errors. With such a range of magnitudes, one must be especially careful to distinguish between average values and maximum or minimum values.

Again it is appropriate to mention the scaling of the problem. If λ_M, the largest of the eigenvalues, is of order unity, it implies that a unitary function

$f(x)$ produces a $\mathbf{g}$ with a norm of approximately unity; if the eigenvalue λ_M is much larger than unity, it implies that either the kernels or the units used are such that the elements of $\mathbf{g}$ tend to be numerically very much larger than $f(x)$. This scaling or mapping aspects of the eigenvalues can be illustrated in several ways. For example, if:

$$g_i = \int K_i(x)\, f(x)\, \mathrm{d}x$$

the Schwarz inequality gives:

$$|g_i|^2 \leqslant \int f^2(x)\, \mathrm{d}x \int K_i^2(x)\, \mathrm{d}x$$

so:

$$g_i^2 \leqslant \int f^2(x)\, \mathrm{d}x \sum_i c_{ii}$$

c_{ii} being the diagonal elements of the covariance matrix $\mathbf{C}$. $\Sigma_i\, c_{ii}$, the trace of the matrix, equals the sum of the eigenvalues, so this sum is an upper bound for the ratio of the square norm $|\mathbf{g}|^2$ to the square norm $\int f^2(x)\, \mathrm{d}x$.

Alternately if $k_i(x)$ and $f(x)$ are expanded in terms of the orthogonal functions $\hat{\phi}_1(x)$, $\hat{\phi}_2(x)$, etc., so that:

$$\mathbf{k}(x) = \mathbf{U}\,\mathbf{\Lambda}^{1/2}\hat{\boldsymbol{\phi}}(x)$$

$$f(x) = \boldsymbol{\xi}^*\hat{\boldsymbol{\phi}}(x) = \hat{\boldsymbol{\phi}}^*(x)\,\boldsymbol{\xi}$$

then obviously:

$$\mathbf{g} = \mathbf{U}\mathbf{\Lambda}^{1/2}\boldsymbol{\xi}$$

Hence:

$$|\mathbf{g}|^2 = \boldsymbol{\xi}\,\mathbf{\Lambda}^{1/2}\mathbf{U}^*\mathbf{U}\,\mathbf{\Lambda}^{1/2}\boldsymbol{\xi}$$

$$= \boldsymbol{\xi}^*\mathbf{\Lambda}\,\boldsymbol{\xi} = \sum_i \lambda_i \xi_i^2$$

and

$$|f|^2 = \boldsymbol{\xi}^*\boldsymbol{\xi} = \sum_i \xi_i^2$$

So:

$$\frac{|\mathbf{g}|^2}{|f|^2} = \frac{\lambda_1\xi_1^2 + \lambda_2\xi_2^2 + \ldots + \lambda_M\xi_M^2}{\xi_1^2 + \xi_2^2 + \ldots + \xi_M^2} \qquad [8.14]$$

which lies between λ_1 and λ_M, and if λ_M is appreciably greater than the other eigenvalues then on average $|\mathbf{g}|^2/|f|^2$ is of the order of λ_M.

Returning now to the main question, that of error magnification, it will be assumed that the problem has been scaled so that $|f(x)|^2 = 1$. The relative error in $f(x)$ is given by [8.13] as:

$$\frac{|\delta(x)|}{|f(x)|} = |\delta(x)| = \sqrt{\lambda_1^{-1}\epsilon_1^2 + \lambda_2^{-1}\epsilon_2^2 + \ldots + \lambda_M^{-1}\epsilon_M^2}$$

At this point the primes have been dropped on ϵ_1, ϵ_2, etc. — i.e. we assume random, uncorrelated errors with r.m.s. values $M^{-1/2}\ |\epsilon|$; the relative error in $f(x)$ will then average $M^{-1/2}\sqrt{(\lambda_1^{-1} + \lambda_2^{-1} + \ldots, + \lambda_M^{-1})}|\epsilon|$. The error in $\boldsymbol{g}$ is ϵ, while the square norm $|\boldsymbol{g}|^2$ is simply $\Sigma_i\ \xi_i^2\ \lambda_i$; thus the relative error in $\boldsymbol{g}$ is:

$$\frac{(\epsilon_1^2 + \epsilon_2^2 + \ldots + \epsilon_M^2)^{1/2}}{(\lambda_1\xi_1^2 + \lambda_2\xi_2^2 + \ldots + \lambda_M\xi_M^2)^{1/2}}$$

which lies between $\lambda_1^{-1/2}\ |\epsilon|$ and $\lambda_M^{-1/2}|\epsilon|$ and (if all directions of $\boldsymbol{\xi}$'s in the vector space are equally probable) averages $M^{-1/2}(\lambda_1 + \lambda_2 + \ldots)^{-1/2}|\epsilon|$.

The magnification of the relative error, i.e. the ratio of the percent error in $f(x)$ to that in $\boldsymbol{g}$ is especially important. It is given by:

$$\sqrt{\frac{|\delta(x)|^2\ |g|^2}{|f(x)|^2\ |\epsilon|^2}} = \sqrt{\frac{(\lambda_1^{-1}\epsilon_1^2 + \lambda_2^{-1}\epsilon_2^2 + \ldots)(\lambda_1\xi_1^2 + \lambda_2\xi_2^2 + \ldots)}{\epsilon_1^2 + \epsilon_2^2 + \ldots}} \qquad [8.15]$$

when the errors are uncorrelated and of equal expected magnitude, then inserting the average value $\epsilon_1 = \epsilon_2 = \ldots = |\epsilon|M^{-1/2}$ and $\xi_1 = \xi_2 = \ldots = M^{-1/2}$ into the above expression one obtains $\sqrt{\Sigma_i\ \lambda_i^{-1}\ \Sigma_i\ \lambda_i}/M$ for this measure of error magnification. The best- and worst-case error magnifications are obtained by setting $\epsilon_M^2 = |\epsilon|^2$, $\xi_1^2 = 1$ and $\epsilon_1^2 = |\epsilon|^2$, $\xi_M^2 = 1$ respectively, giving:

best-case relative error magnification — $\sqrt{\lambda_1/\lambda_M}$ or $\sqrt{\lambda_{min}/\lambda_{max}}$

worst-case relative error magnification — $\sqrt{\lambda_M/\lambda_1}$ or $\sqrt{\lambda_{max}/\lambda_{min}}$

When the largest eigenvalue λ_M dominates the sum $\Sigma_i\ \lambda_i$ and the smallest λ_1 dominates the sum $\Sigma_i\ \lambda_i^{-1}$, the average relative error magnification can be written simply as $M^{-1}\ \sqrt{\lambda_M/\lambda_1}$, or M^{-1} times the worst-case value.

Information content. In a given M measurements one should clearly exclude from the solution any components for which the error magnification is excessive. Just exactly what is excessive and which error magnification — average or worst-case — is appropriate will depend on circumstances. If we use the average relative error magnification calculated above and delete all components for which the average error exceeds the expected average value of that component, the criterion for rejection in the presence of relative error

r is

$$(r/M)\sqrt{\lambda_M/\lambda_k} > 1$$

or

$$\lambda_k/\lambda_M < (r/M)^2$$

Since a ratio of unity is in fact a very marginal situation, the criterion is better written as:

$$\lambda_k/\lambda_M \ll (r/M)^2$$

or, unless M is quite large:

$$\lambda_k/\lambda_M \ll r^2 \qquad [8.16]$$

a criterion that is safe, but not excessively so.

The information content of M measurements must therefore be reduced by one *for each eigenvalue* of the kernel covariance matrix which falls below the threshold of acceptability specified by equation [8.16].

BIBLIOGRAPHY

Courant, R. and Hilbert, D., 1953. *Methods of Mathematical Physics, I.* Interscience, New York, N.Y., 561 pp.

Deutsch, R., 1965. *Estimation Theory.* Prentice-Hall, Englewood Cliffs, N.J., 269 pp.

Mateer, C.L., 1965. On the information content of Umkehr observations. *J. Atmos. Sci.*, 22: 370—381.

Strand, O.N. and Westwater, E.R., 1968. Statistical information content of radiation measurements used in indirect sensing. *J. Atmos. Sci.*, 25: 750—758.

Twomey, S., 1966. Indirect measurements of atmospheric temperature profiles from satellites, II. Mathematical aspects of the inversion problem. *Mon. Weather Rev.*, 94: 363—366.

CHAPTER 9

Further topics

The inversions which have been done up to now all have related to a fictitious example involving as kernels the functions $x \exp(-y_i x)$ ($i = 1, 2 ..., M$). The results obtained from the inversion procedures were in many cases disappointing, inasmuch that the mathematical manipulations usually failed to produce inversions that were very close to the original $f(x)$ unless constraints were applied which strongly forced the hand of the inversion scheme. Yet the residuals, as measured by the number $\Sigma_i\ [g_i - \int K_i(x)\, f'(x)\, dx]^2$, were individually and collectively small when these less than satisfactory solutions were inserted for $f'(x)$ in the integral, and that is the most important part of the exercise. We have by different methods obtained various inversions all of which were satisfactory when judged by the residuals — and in a practical field measurement situation that is the only criterion of a satisfactory solution and a successful inversion in most situations.

Inversions can and do give better results than those which have been given here, but only because either (1) the kernels involved are better, from the viewpoint of their mutual independence and high-frequency content, than the set of kernels chosen for illustration, or (2) a sufficient amount of a priori information existed about the behavior of $f(x)$. In a hypothetical experiment one can choose any set of kernels and in so doing one is limited only by one's imagination. Perhaps not a great deal of imagination was exercised in the choice of the kernels $x\ e^{-yx}$ and much more appealing inversions could have been obtained by a more imaginative choice. However, the kernels of most real physical inversion problems are similar in character to those chosen, being predominantly exponential in character. We do not find, for example, optical transmissions which remain close to unity well down through the atmosphere and change to a zero value over a distance which is short compared to that over which they remained constant or almost so: radiation which is not quickly attenuated can penetrate well through a considerable depth of atmosphere but ipso facto it will not change quickly within a thin layer. It is for this reason that most inversions illustrated up to now gave somewhat better results at small values of x than they did at large values. To obtain an exponential kernel which changes rapidly with x it is evidently necessary that y be large; such a kernel, however, will also become quite small as x increases. Conversely an exponential which remains appreciable out to large values of x necessarily changes slowly with x. This kind of behavior is a physical necessity, even if it is undesirable from the point of view of inversion mathematics.

9.1. FURTHER EXAMPLES OF INVERSIONS AND THEIR BEHAVIOR

When something is known a priori about the $f(x)$ which is being sought, beyond the fact that it is non-negative, a great improvement can often be made in the solution by applying additional constraints utilizing this a priori information. In cases such as ozone distribution or particle size inference by inversion a value for $\int_a^b f(x)\,dx$ is known (the total ozone in the column or the total number of particles) or can be easily determined directly. In other inversion problems one or both end-point values of $f(x)$ are pre-set by the physical nature of the problem. A few examples will be given to show the application of constraints in this way and to demonstrate that an improvement in the quality of the inversion solutions are thereby produced.

(1) The constrained linear inversion according to equation [6.4] was applied to the standard problem, using the same data and number of data points in g_i as were employed previously, but the end points $f(0)$ and $f(1)$ were taken to be known a priori and to be $f(0) = f(1) = 2$. This constraint was then incorporated into the constraint matrix **H** as described in Chapter 6. The smoothness measure which was used was the sum of squares of second differences. The result of the inversion (Fig. 9.1) is much closer to the initial "true" $f(x)$ and furthermore was found to possess more stability towards errors in $\mathbf{g}$ — a given value of γ produced smaller residuals.

(2) A second example of an inversion which is much more encouraging than some of the early examples relates to the problem of inferring the size distribution of aerosol particles from measurements of the number of particles remaining after the aerosol has been passed through a set of filters

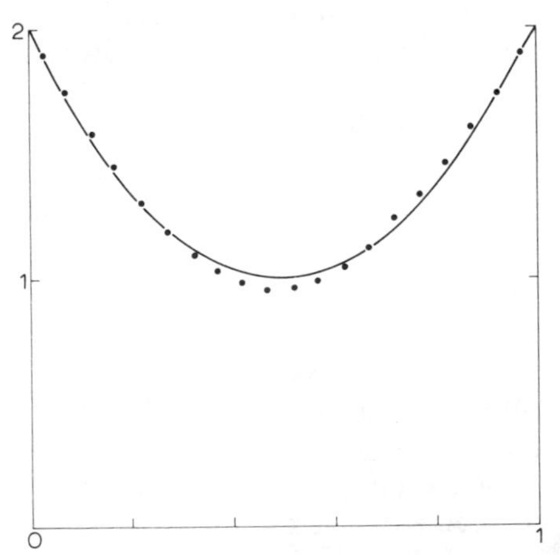

Fig. 9.1. Inversion of the "standard problem" using constrained endpoints; this would be possible in some real inversion situations where the endpoint values of the unknown $f(x)$ are known a priori.

the transmissions of which are shown in Fig. 9.2. Starting with an initial $f_0(x)$ and a set of values of g_i computed for a hypothetical $f(x)$, the non-linear iterative procedure of equation [7.21] was then applied. The solution obtained after 100 cycles of iteration — 2000 individual iterative adjustment steps — is shown in Fig. 9.3 for the flat first guess $f_0(x)$ shown dotted in the figure. The open circles in the figure represent the solution obtained with a drastically different $f_0(x)$. The solutions — solid and open circles — differ very little over most of the range of x from the "true" $f(x)$, which is the heavy line in the figure. The "right" answer is recovered in this example to a very satisfactory degree of fidelity, the position and height of the peak and the slope of the unknown distribution being reproduced closely; the very steep fall-off of the assumed distribution to the left of the peak was not followed as closely — that implies a high-frequency component in $f(x)$ which could not be retrieved, but it did not produce any troublesome oscillatory features in the solution. This stability was confirmed by the deliberate introductoin of errors into $\mathbf{g}$ before inversion was commenced. The result was a solution which was still very good. Table 9.1 shows for the most significant portion of the distribution the values of the initial distribution $f(x)$, the solution by inversion without errors (other than computer roundoff) and the solution when deliberate additional errors (which were $\pm 4\%$ with r.m.s. value 2.3%) were included. The failure of these errors to greatly affect the solu-

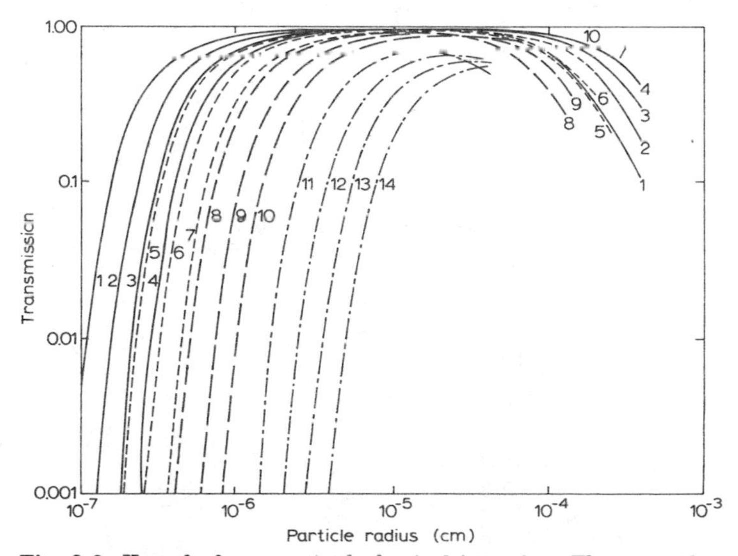

Fig. 9.2. Kernels for an actual physical inversion. These are the transmission curves for polycarbonate (Nuclepore) filters. Changing the air flow rate or the physical dimensions of the filters changes the transmission. The x-scale is the logarithm of the radius of particle in centimeters.

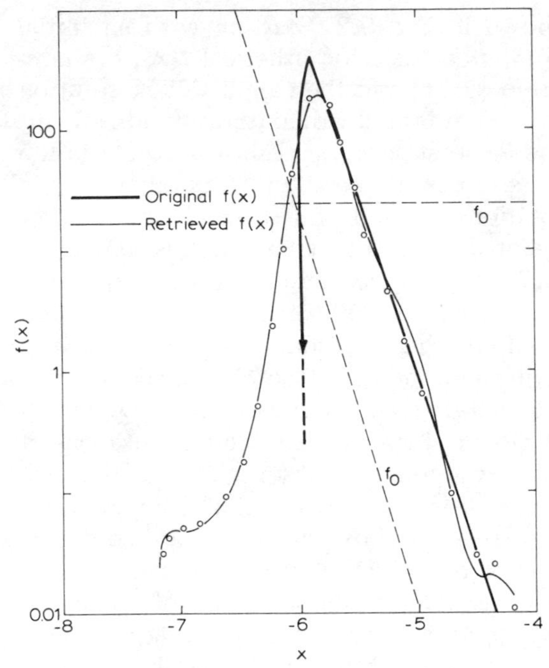

Fig. 9.3. Results of inversion by the non-linear iterative scheme described in Chapter 7, using *calculated* data for the number of particles surviving passage through the set of filters for which the transmission data was given in Fig. 9.2. The sloping first-guess $f_0(x)$ was used for the results plotted as open circles, the horizontal $f_0(x)$ gave the solution drawn as a light solid line.

TABLE 9.1

Inversion of aerosol size-distribution problem using iterative non-linear inversion

x	$f(x)$	$f'(x)$	$f'(x)$ with errors added
−6.1	0	1.84	2.4
−6.0	0	8.6	10.0
−5.9	32.4	30.0	31.9
−5.8	73.1	68.9	70.8
−5.7	125.0	95.0	94.4
−5.6	63.1	75.4	72.6
−5.5	31.6	38.5	37.3
−5.4	15.9	16.0	16.2
−5.3	7.94	6.7	7.0
−5.2	3.98	3.0	3.2
−5.1	1.96	1.5	1.6
−5.0	1.0	0.8	0.8

tion is noteworthy and more than a little surprising. In connection with the relatively small effect of the added errors it should be borne in mind that the procedure followed — that of equation [7.21] — is virtually incapable of manufacturing highly oscillatory components in $f(x)$. The addition of an arbitrary random error into the elements of $\mathbf{g}$ can be regarded as that of three components:

(a) An error component which is outside the vector space spanned by vectors which are related to real positive $f(x)$ through the fundamental transform $\mathbf{g} = \int \mathbf{k}(x)\, f(x)\, \mathrm{d}x$.

(b) An error component which is *outside* the vector space containing all vectors which through the integral transform map $f(x)$'s which can be generated by the iteration procedure which was used. That vector space is a subspace of the vector space of (a).

(c) An error component which lies *in* the vector space of (b). Only this component can propagate errors into the inversion $f'(x)$.

The behavior of the residuals is also illuminating. Fig. 9.4 shows the r.m.s. and maximum residuals, i.e. $(1/M)\ \Sigma_i\ (g_i' - g_i)^2/g_i^2$ and $\max |(g_i' - g_i)|/g_i$, plotted against the number of cycles of iteration. The r.m.s. residual decreased to a small value (slightly over 1%) when the added error was not present, but in the presence of the additional error the r.m.s. residual was 3.3% after fifty iteration cycles and 2.4% after one hundred cycles. The latter is very close to the r.m.s. value of the added error (2.3%), and suggests a measure of discrimination in the inversion algorithm against random error, in the sense that it finds difficulty in inverting such an error component. This is a very useful property, but a rigorous mathematical explanation or justifica-

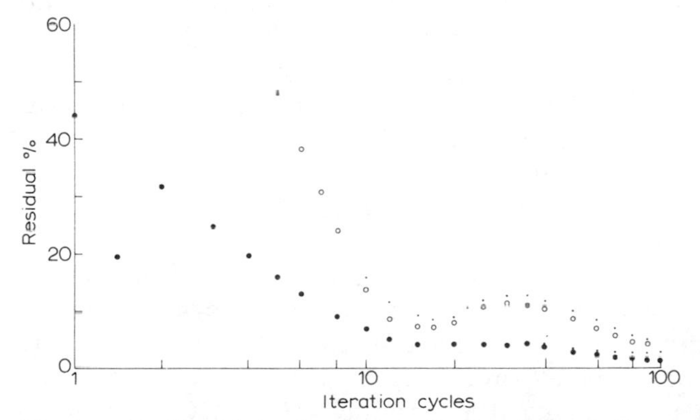

Fig. 9.4. Residual behavior during the iterative solution process. Open circles refer to an iteration starting with a very poor first-guess solution. Solid symbols refer to a first-guess which was reasonably realistic. Upper (small) dots represent results obtained when additional error (r.m.s. 2.7%) was introduced into the simulated measurements before inversion iteration was begun.

tion for it does not exist. The comments made above about the several possible components which can exist in a random error and the differences which these components, (a), (b) and (c), exhibit with respect to a particular integral transform, make it appear at least plausible that a measure of error rejection is "built into" inversion algorithms of this kind. The writer's experience with non-linear iterative procedures of this sort supports this possibility; they seem to be quite generally more stable than most other methods. Clearly, however, a considerable amount of numerical experimenting is called for in any given inversion problem; the experiments should be done using the kernels which one intends to utilize when measurements eventually will be inverted. Only by such numerical experiments can the numerical analyst obtain a real appreciation of the strengths and weaknesses of the inversion procedure. At the time of writing our theoretical knowledge of linear algorithms (whether iterative or non-iterative) is reasonably sound, but this is not true for non-linear algorithms, and our approach there must necessarily be more tentative — the procedures are primarily heuristic and our insight into their properties are at best qualitative and intuitive. They are nonetheless of considerable value and power.

The result of the application of the methods just discussed to the inversion of an actual measurement of aerosol particle concentrations emerging from a set of filters with transmission functions those plotted in Fig. 9.2 is given in Fig. 9.5. Again the solution is shown for two first guesses differing widely from each other, and it is seen that the solution was quite insenstivie to the first guess from which the iteration was started.

Problems in which the object function $f(x)$ — the retrieval of which is the goal of the inversion — is likely to extend over a wide "dynamic range" are not readily handled by the methods of Chapter 6. The solution given in Fig. 9.5 was of this character and so was the hypothetical $f(x)$ of Table 9.1 and Fig. 9.3: they all require the use of a logarithmic scale to display them usefully, and though smooth as presented on that scale, they are not smooth on a linear scale, as can be easily verified. The difficulty of inversion by linear methods arises primarily from this non-smoothness. A function which is linear on the logarithmic scale used in Fig. 9.5 would represent a power distribution of the form $f(x) = \text{constant} \cdot x^{-\alpha}$ and transforming the problem in $f(x)$ to one in $x^{\alpha} f(x)$ would remove the difficulty. The slope parameter α, however, is not known a priori and use of an arbitrary value of α represents a very great distortion if the true slope of $f(x)$ on a logarithmic plot happens to be seriously different to the value used for α. Non-linear iterative methods on the other hand do not "force" any particular slope in the solution; they cannot of course by their nature produce arbitrarily steep slopes in the solution $f(x)$. The behavior of the solution in Fig. 9.3 in the region to the left of the maximum suggests that the non-linear iteration scheme and the kernels of Fig. 9.2 can produce slopes down to about seven decades per decade in that region of x; the limit will of course be different in regions

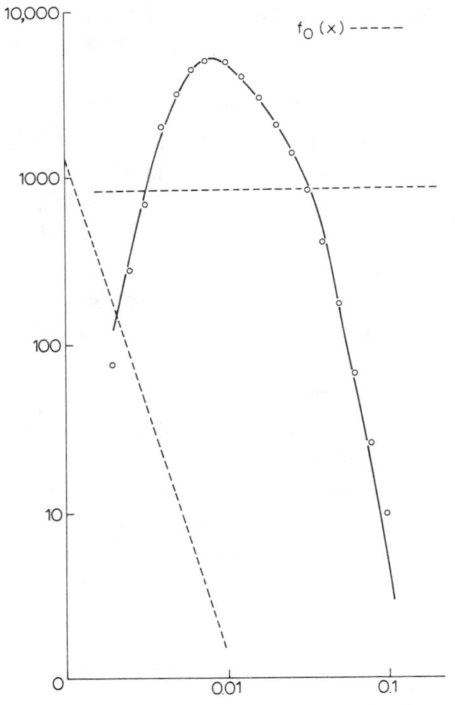

Fig. 9.5. Solution obtained by the iterative algorithm applied to *experimentally measured* data. The curve relates to the sloping first guess, the open circles show the result of starting with the horizontal line as first guess. Broken curves are the first guesses.

where the kernels are different in their behavior and it will be different for a different set of kernels. It will also be influenced by the number by iteration cycles.

9.2. BENEFICIAL ASPECTS OF KERNEL INTERDEPENDENCE

Throughout the preceding chapters repeated allusion has been made — always in an adverse vein — to the lack of independence existing among the measurements of g_i ($i = 1, 2, ..., M$) in most indirect sensing problems. This lack of independence, caused by correlations between the kernels $K_1(x)$, $K_2(x)$, ..., $K_M(x)$, has been shown to be the root cause for instability, near-singular matrices and systems of equations, and other weaknesses that inversions are heir to. In the preceding chapter, there were derived and applied explicit formulae which provided linear relationships between the various g_i; in many cases these linear relationships enabled some of a set of measured g_i to be predicted from the others to better accuracy than the best measurement could provide. Examples of the predictability of measured quantities were also given — Mateer's (see Chapter 8) analysis of the interdependence of

Umkehr measurements provided a rather spectacular demonstration of the very high degree of interdependence often found in real physical measurements.

The proverb that it is an ill wind that blows no good finds realization in this context, for the existence of strong correlations among the g_i has at least one beneficial effect: a bad measurement or a faulty prescription of a single kernel or even a few kernels can usually be diagnosed from the results of an inversion. That of course would not be possible if the kernels were strictly independent, since then each measurement would also be completely independent of all others. In the presence of correlations there are inter-relationships among the individual measurements and a given measurement must "fit in" with the group as a whole.

In discussing iterative methods the identification of error levels by the flattening-out of the curve of diminution of the residual was commented on and illustrated. That behavior is also an example of the ability of some inversion algorithms to "know" something about the errors in the measurements which are being inverted. Even more direct examples can be found, both with iterative non-linear and direct linear methods. The constrained linear inversion formulae of Chapter 6 point out a bad measurement by using up a disproportionate amount of the prescribed residual norm $|\epsilon|^2$ in a single ϵ_k; the suspicion that that particular g_k is in some way anomalous can often be confirmed by deleting it from the vector $\boldsymbol{g}$ and deleting the kth kernel from the array of kernels, obtaining a $(M - 1 \times N)$ system which is then solved anew. If the result is a more uniform assignment of the magnitudes of the ϵ_i (with, for example, approximately equal magnitudes for positive and negative ϵ) one can be reasonably confident that the measured g_k which was deleted was in some way defective. If at the same time the second solution is also smoother than the one first obtained (as indicated by the value of $\boldsymbol{f}^*\mathbf{H}\boldsymbol{f}$, or by plotting the solutions), this provides further evidence that the relation $g_k = \int K_k(x)\, f(x)\, \mathrm{d}x$ was contradictory.

This is illustrated by the following example. The g_i of the standard problem were left unchanged except for the last, g_{20}, which was reduced about 10%. In Table 9.2 the individual residuals ϵ_i — calculated as elements of the vector $(\boldsymbol{g} - \mathbf{A}\boldsymbol{f})$ — are shown for the problem with and without the deliberate pollution of the last element of $\boldsymbol{g}$. The solution was obtained using equation [6.4] with $\mathbf{H}$ in the form appropriate to the second-difference measure Σ_2^{N-1} $(f_{j-1} - 2f_j + f_{j+1})^2$; a random error of $\pm 0.5\%$ was introduced in the g_i to simulate the situation existing when a measurement of good quality was the source of the g_i.

The difference between the situations with and without a large added error in g_{20} is obvious; for a given γ — the value to which the table corresponds being 10^{-3} — the residuals ϵ_i were distributed more or less randomly about zero when the errors introduced into $\boldsymbol{g}$ were small and random, but when the individual error in g_{20} was anomalously large, the residuals were all

TABLE 9.2

Solution residuals with and without large added error introduced in 20th measurement

i	ϵ_i (small random error only)	ϵ_i (added error)
1	+1	+1
2	0	+5
3	−1	+13
4	−1	+18
5	−2	+20
6	+2	+18
7	0	+15
8	0	+10
9	−1	+4
10	0	−3
11	+6	−10
12	−8	−16
13	−1	−22
14	+12	−29
15	+8	−25
16	−1	−41
17	−16	−46
18	−12	−50
19	−12	−56
20	−1	+240
$\lvert\epsilon\rvert^2$	868	71557
$f^* \mathrm{H} f$	6.3×10^{-3}	1.6×10^{-2}

of the same positive sign up to $i = 9$, then all the same sign but negative up to $i = 19$, with a single very large positive residual for $i = 20$. Such a table of residuals very clearly points to g_{20} as the offender. In this particular instance the result of the added large error was not serious so far as the solution $f(x)$ was concerned. Table 9.3 shows the values obtained for $f(x_i)$ with and without the added error in g_{20}. The possibility of errors being recognized as such depends on correlations among the measurements. The particular error used for the example just given involved g_{20} being altered to 0.3122 from its correct value 0.3422 and a visual inspection of the array $\boldsymbol{g}$ would have suggested that g_{20} was in error, for the diminishing sequence of g_i's down to $i = 19$ was followed by an increase between g_{19} and g_{20} (the values used were those listed in Table 6.3, so that the last four values were 0.3063, 0.3200, 0.3318, 0.3122), but contradictory values of g_i usually could not be detected by mere inspection.

If in a real experimental situation one particular residual is notably larger than the others, it most probably implies that the measurement of that particular g_i was made with a larger error than the rest just through the

TABLE 9.3

Solution with and without added error in g_{20}

j	$f(x_j)$ without added error in g_{20}	$f(x_j)$ with added error in g_{20}	$f(x_j)$ g_{20} deleted
1	1.9025	1.9020	1.9025
2	1.7440	1.7400	1.7433
3	1.5906	1.5933	1.5914
4	1.4454	1.4715	1.4557
5	1.3128	1.3768	1.3200
6	1.1968	1.3056	1.1992
7	1.1010	1.2514	1.1008
8	1.0278	1.2065	1.0303
9	0.9800	1.1646	0.9916
10	0.9576	1.1221	0.9674
11	0.9622	1.0791	0.9605
12	0.9929	1.0395	0.9888
13	1.0485	1.0113	1.0032
14	1.1270	1.0050	1.1243
15	1.2256	1.0325	1.1987
16	1.3412	1.1048	1.3029
17	1.4703	1.2303	1.4755
18	1.6092	1.4112	1.6454
19	1.7544	1.6413	1.7719
20	1.9025	1.9024	1.9025

natural random behavior of measurement errors. If, however, the same residual is consistently large and has the same sign from one group of measurements to the next, this indicates that that particular measurement is defective; it is not always the experimental accuracy that is at fault — it may also be the kernel associated with the measurement. In more than one instance in the writer's experience the residual behavior on inversion has been the first indication that a kernel was incorrect. This can arise because of a faulty prescription of physical variables involved, as occurred in one series of experiments when a manufacturer's specification for a filter was incorrect. The source of error may also be more subtle, as the following account of an actual experience will show.

In an aerosol sizing experiment, samples of an aerosol stored on a large container were passed into a particle counter through a "battery" of accurate shallow ducts of rectangular cross-section; the aerosol could be passed through the duct system once or many times and the emerging particle number concentration was measured after 1, 2, 3, 4, 5, 7 and 14 passages through the system. Diffusion theory shows that the fraction of aerosol particles which emerges from a rectangular channel is related to the diffusion coefficient D — a known function of size — by:

$$\frac{N}{N_0} = c_1 \exp(-a_1 Dl/q) + c_2 \exp(-a_2 Dl/q) + c_3 \exp(-a_3 Dl/q) + \ldots$$

an infinite series in which the sum of the coefficients c_i must evidently be 1 when the series $c_1 + c_2 + c_3 + \ldots$, is summed to infinity, since N/N_0 must be unity for a channel of length $l = 0$ or an infinite flow rate q. The c_i are all positive and to obtain a partial sum exceeding 0.99 it was found that six terms were requried. The parameters a_i in the above equation involve the physical dimensions of the channel cross-section and a characteristic number which increases with increasing i. The measurements were used as input to an inversion procedure to obtain a particle size distribution $N_0\ f(x)$. If N_i/N_0 — which is derived directly from the measured quantities — is equated to g_i, one can write:

$$g_i = \int_0^\infty K_i(x) f(x)\,dx$$

if $K_i(x)$ is defined to be $\{\Sigma_i\ c_i \exp(-a_i\ D(x)\ l\ q^{-1})\}^{p_i}$, p_i being the number of passes through the channel for the ith measurement. When the equation was solved for $f(x)$ by the methods of Chapter 6 it was noticed that most of the norm $|\epsilon|^2$ was used up for the measurements involving the greatest number of passes through the channels. This occurred systematically over many experiments. When the "message" was eventually interpreted, it was realized that even though six terms sufficed to give N/N_0 to better than 1% for single pass through the channel, a greater number of terms was needed to account for multiple passes, since $\{\Sigma_i^n\ c_i\}^p$ is smaller than $\Sigma_i^n\ c_i$ except when the sum is *exactly* unity: raised to the seventh power 0.99 becomes 0.932 and to the fourteenth power it is 0.869. When additional coefficients of the series were computed so that $\{\Sigma_i\ c_i\}^p \geqslant 0.99$ for $p \leqslant 14$, the predilection of the algorithm for placing most of the assigned norm $|\epsilon|^2$ into positions associated with high numbers of passes through the channel system was eliminated; the solutions became smoother and the residuals ceased to exhibit highly systematic behavior.

By not retaining a sufficient number of terms in the series defining the kernels the magnitude of the kernels was underestimated, the effect being more serious the greater the number of passes involved in the measurement. Thus instead of the relation:

$$\mathbf{A}f = g$$

a relation of the form $\mathbf{A}'f = g$ was being used with $\mathbf{A}'$ being $\mathbf{DA}$, $\mathbf{D}$ a distortion of the identity matrix, in which the lower elements decrease further below unity with increasing index i. The solution f as a result tends to $(\mathbf{A}')^{-1}$ g, i.e. to $\mathbf{A}^{-1}\mathbf{D}^{-1}\ g$. The constraint tends partially to hold it in the direction of $A^{-1}g$ and the residual can be written as:

$$\epsilon_D = g - DA[A^{-1}D^{-1}g(1-\beta) + \beta A^{-1}g]$$

$$= \beta(I - D)g$$

$$= \beta \begin{bmatrix} 0 & & & & & \\ & 0 & & & & \\ & & \cdot & & & \\ & & & \cdot & & \\ & & & & \xi_{m-1} & \\ & & & & & \xi_m \end{bmatrix} g$$

ξ_i represents the deviation below unity of those elements of **D** which are seriously different from unity. β is a small positive number which will depend on the constraint parameter γ. The residual vector contains a component which increases with increasing index when the distortion matrix **D** is a diagonal matrix with elements which start close to unity but begin to deviate increasingly from it as one proceeds down the diagonal.

9.3. INFERENCE OF MORE UNKNOWNS THAN THERE ARE MEASUREMENTS

In conventional algebra one can obtain no more than M unknowns from a system of M equations. Earlier in this chapter and elsewhere examples have cropped up where this principle is apparently being violated. In Table 9.3, for example, the solution in the last column — that for the deliberately added error in g_{20} — was found to be improved when the twentieth measurement was deleted, yet that left a system of 19 linear relations which had to be solved for twenty unknowns. Obviously no real contradiction is involved. A constrained inversion with a (19×20) matrix for **A** is quite different to an unconstrained system of equations. The additional degree of freedom is absorbed by the requirement that $f^*\mathbf{H}f$ be minimized. From a vector space viewpoint a system of 19 equations in 20 unknowns would be satisfied by all points of a line in vector space; the quantity $f^*\mathbf{H}f$ will vary along this line and the constrained inversion can find a unique solution, since it minimizes that quantity. The twenty unknowns which the constrained solution returns are not of course twenty truly independent quantities since the application of constraints introduces correlations among the unknowns $f(x_i)$, reducing the real dimension of the problem from N to $N - L$ where L is the number of eigenvalues less than γ. It is for this reason that a system of M measurements can be successfully solved for more than M "unknowns".

9.4. INVERSIONS IN WHICH THE UNKNOWN IS A MATRIX

The problem of determining or estimating f, given $g = Af$ has an analogue in the problem — in a sense the inverse problem — of determining A, given a set of g's and f's. That this is possible when at least N such pairs are given (N being the dimension of a square matrix A) is easily demonstrated by the very special case:

$$\lambda_1 u_1 = Au_1$$

$$\lambda_2 u_2 = Au_2$$

$$\vdots$$

$$\lambda_N u_N = Au_N$$

with the vectors u mutually orthogonal; the solution for this special case is of course:

$$A = U\Lambda U^* = \| u_{ik} \lambda_k u_{jk} \|$$

If N independent vectors $f_1, f_2, ..., f_N$, are given together with N vectors $g_1, g_2, ..., g_N$, such that:

$$g_k = Af_k \qquad (k = 1, 2, ..., N)$$

then it is a straightforward matter to obtain an equation for A. All that is required is the postmultiplication of each of these equations (which are vector equations) by f_k^*; this gives an equation in which all the factors are matrices:

$$g_k f_k^* = Af_k f_k^*$$

Clearly one cannot obtain A uniquely with a single value of k, but if N independent f's and the corresponding g's are given, a soluble matrix equation is obtained by summing the last equation over all k. The result is:

$$A = XC^{-1} \qquad [9.1]$$

C being a covariance matrix or correlation matrix for the f's, $\langle f_k f_k^* \rangle$, while X is a cross-correlation matrix $\langle g_k f_k^* \rangle$.

If the set f_k ($k = 1, 2, ..., N$) does not span the N-dimensioned vector space it implies that the set is not made up of independent vectors; the requirement of independence assures that C will not be singular and it is therefore sufficient and necessary.

The problem of finding A can be course be solved also by finding the N vectors which comprise the columns of A. The equation:

$$X = AC$$

transposed and postmultiplied by the vector $\boldsymbol{e}_k$ gives:

$$\mathbf{X}^* \boldsymbol{e}_k = \mathbf{C}^* \mathbf{A}^* \boldsymbol{e}_k$$

but postmultiplication of a matrix by $\boldsymbol{e}_k$ produces a vector which is the kth column of the matrix. Writing $\boldsymbol{a}_k$ for the kth row of $\mathbf{A}$, it is evidently given by the equation:

$$\boldsymbol{a}_k = (\mathbf{C}^*)^{-1} \boldsymbol{x}_k \qquad [9.2]$$

$\boldsymbol{x}_k$ being of course the kth row of the cross-correlation matrix $\mathbf{X}$. If experimental or inexact data are involved, direct inversion according to equations [9.1] or [9.2] will not be satisfactory unless the eigenvalues of $\mathbf{C}$ are all large — which they will be if the f's are mutually orthogonal or close to it. Otherwise a constrained solution for the rows of $\mathbf{A}$ (the columns of $\mathbf{A}^*$) can be attempted, writing:

$$\mathbf{C}^* \boldsymbol{a}_k = \boldsymbol{x}_k$$

and the constrained solution in the usual way is:

$$\boldsymbol{a}_k = (\mathbf{C}\mathbf{C}^* + \mathbf{H})^{-1} \mathbf{C} \boldsymbol{x}_k \qquad [9.3]$$

An alternative procedure is through a synthesis approach similar to that involved in the Gilbert-Backus inversion method. If $\boldsymbol{f}_k$ were the special vector $\boldsymbol{e}_k$ the equation $\boldsymbol{g} = \mathbf{A}\boldsymbol{f}$ would give explicitly the kth column of $\mathbf{A}$. If therefore a formula can be obtained such that for f's which are not these special vectors one can write:

$$\sum_j \xi_{lj} \boldsymbol{f}_j \cong \boldsymbol{e}_l$$

then the lth column of $\mathbf{A}$ is given to a similar degree of approximation by the combination $\Sigma_j \, \xi_{lj} \, \boldsymbol{g}_j$.

9.5. PREDICTION AND INVERSION

The problem of predicting x_{n+1} given $x_1, x_2, ..., x_n$ is not in the usual sense an inversion problem; in a strict mathematical sense it is not a soluble one. Nevertheless, a large number of numerical exercises in a wide variety of disciplines from economics to population studies and flood control are devoted to the question of inferring x_{n+1} from the preceding values of x, at least in a probabilistic sense. Without any knowledge of the basic causes for the evolution of x through its previous values there are strictly no grounds even for asserting that x_{n+1} will be positive when all the x_i up to $i = n$ have in fact been positive. On the other hand, if x is a random variable which is normally distributed the probability that x_{n+1} will be positive when x_i (i =

1, 2, ..., n) have been positive can be calculated and of course if n is large the calculated probability will be high.

In quite a number of contexts an approach to prediction has been followed in which a linear relationship between the x_i is envisaged to exist. (Gabor once proposed a learning algorithm in which non-linear combinations of the x_i were also considered.) Intuitively it seems reasonable to suppose that a very remote past value $x_k (k << n)$ is less relevant to the determination of x_{n+1} than the recent values x_n, x_{n-1}, etc. This suggests the possibility of using a linear relationship of the form:

$$x_n = \xi_1 x_{n-1} + \xi_2 x_{n-2} + \ldots + \xi_p x_{n-p}$$

for predicting a future value of the quantity x, given the present value and a sufficiently long sequence of past values. To make such a formula of any predictive value it is clearly necessary that a similar relationship relate x_{n-1} to x_{n-2}, x_{n-3}, ..., x_{n-p-1}, and so on. The set of coefficients should be the same, or at least should change only slowly, with n. That implies that similar relationships extend back into the past data, so that:

$$\left.\begin{array}{l} x_n = \xi_1 x_{n-1} + \xi_2 x_{n-2} + \ldots + \xi_p x_{n-p} \\ x_{n-1} = \xi_1 x_{n-2} + \xi_2 x_{n-3} + \ldots + \xi_p x_{n-p-1} \\ x_{n-2} = \xi_1 x_{n-3} + \xi_2 x_{n-4} + \ldots + \xi_p x_{n-p-2} \\ x_{n-3} = \xi_1 x_{n-4} + \xi_2 x_{n-5} + \ldots + \xi_p x_{n-p-3} \\ \\ x_{n-p+1} = \xi_1 x_p + \xi_2 x_{p-1} + \ldots + \xi_p x_1 \end{array}\right\}$$

Provided $n > 2p$ the system of equations could be solved for the unknowns ξ_1, ξ_2, ..., ξ_p in a least squares sense, the result being:

$$\boldsymbol{\xi} = (\mathbf{X}^* \mathbf{X})^{-1} \mathbf{X}^* \boldsymbol{x}$$

where $\mathbf{X}$ and $\boldsymbol{x}$ are respectively:

$$\begin{bmatrix} x_{n-1} & x_{n-2} & x_{n-3} & \cdots & x_{n-p} \\ x_{n-2} & x_{n-3} & x_{n-4} & \cdots & x_{n-p-1} \\ x_{n-3} & x_{n-4} & x_{n-5} & \cdots & x_{n-p-2} \\ \cdot & \cdot & \cdot & & \cdot \\ \cdot & \cdot & \cdot & & \cdot \\ x_p & x_{p-1} & x_{p-2} & & x_1 \end{bmatrix} \text{ and } \begin{bmatrix} x_n \\ x_{n-1} \\ x_{n-2} \\ \cdot \\ \cdot \\ x_{n-p+1} \end{bmatrix}$$

It is evident that if any of the elements of $\boldsymbol{\xi}$ become very large the prediction of x_{n+1} will be unstable, since a very small change in some of the past values will lead to a gross change in the predicted x_{n+1}. It is therefore desirable to seek a constraint to limit the magnitude of the individual elements of $\boldsymbol{\xi}$. The constraint $\boldsymbol{\xi}^*\boldsymbol{\xi} \leqslant c^2$ is obviously suggested, and this leads to the familiar solution:

$$\boldsymbol{\xi} = (\mathbf{X}^*\mathbf{X} + \mathbf{I})^{-1}\mathbf{X}^*\boldsymbol{x}$$

using equation [6.4].

The circumstances under which such procedures are of value can only be assessed by going through the past data and testing the predictive ability of the procedure by evaluating $\boldsymbol{\xi}$ for a sequence of past values of n and comparing the thus predicted x_{n+1} with what actually happened. It is informative, however, to look at the related question: what kind of sequence of x's is generated by repeated application of the predictive formula:

$$x_{p+1} = \xi_1 x_p + \xi_2 x_{p-1} + \ldots + \xi_p x_1$$

That question can readily be answered if vectors $\boldsymbol{x}_1$, $\boldsymbol{x}_2$, etc., are defined as follows:

$$\boldsymbol{x}_1 = \begin{bmatrix} x_p \\ x_{p-1} \\ x_{p-2} \\ \cdot \\ \cdot \\ \cdot \end{bmatrix}, \qquad \boldsymbol{x}_2 = \begin{bmatrix} x_{p+1} \\ x_p \\ x_{p-1} \\ \cdot \\ \cdot \\ \cdot \end{bmatrix}, \qquad \boldsymbol{x}_k = \begin{bmatrix} x_{k+p-1} \\ x_{k+p-2} \\ x_{k+p-3} \\ \cdot \\ \cdot \\ \cdot \end{bmatrix}$$

The relationship $x_{p+1} = \Sigma_k \; \xi_k x_{p+1-k}$ has the consequence that the vectors $\boldsymbol{x}_2$ and $\boldsymbol{x}_1$ are related by:

$$\boldsymbol{x}_2 = \begin{bmatrix} \xi_1 & \xi_2 & \xi_3 \ldots & \xi_p \\ 1 & 0 & 0 \ldots & 0 \\ 0 & 1 & 0 \ldots & 0 \\ \cdot & \cdot & \cdot & \cdot \\ \cdot & \cdot & \cdot & \cdot \\ 0 & 0 & 0 \ldots & 1 \end{bmatrix} \begin{bmatrix} x_p \\ x_{p-1} \\ x_{p-2} \\ \cdot \\ \cdot \\ x_l \end{bmatrix} = \Phi\,\boldsymbol{x}_1$$

And in general:

$$\boldsymbol{x}_k = \Phi\,\boldsymbol{x}_{k-1} = \Phi^2\boldsymbol{x}_{k-2} = \ldots = \Phi^{k-1}\boldsymbol{x}_1$$

The matrix Φ is evidently singular and for certain initial $\boldsymbol{x}_1$ all subsequent

x_k will be zero. Only those eigenvectors which are associated with non-vanishing eigenvalues can give a sequence of non-vanishing x_k. If x_l is such an eigenvector, x_k becomes $\lambda^{k-1}\, x_l$, λ being the associated eigenvalue. If λ is written as $|\lambda|\, e^{i\theta}$, x_k is seen to be:

$$x_k = |\lambda|^{k-1}\, e^{i(k-1)\theta}\, x_l$$

If $|\lambda| < 1$, $x_k \to 0$, while if $|\lambda| > 1$, x_k grows without limit. Only when $|\lambda| = 1$ can a stable sequence of values of x_k be generated. In that event $x_k = x_1\, e^{i(k-1)\theta}$, so it is a periodic function of k with period $2\pi/\theta$.

The sequence of values generated by repeated application of a linear predictive formula of length p must therefore do one of three possible things; it will (1) tend to zero, (2) grow without limit, or (3) be periodic, with at most p discrete frequencies. For a stable set of x's to be generated the set of predicting coefficients $\xi_1, \xi_2, \ldots, \xi_p$ must be such that the roots of the characteristic polynomial are complex numbers of modulus unity. The characteristic polynomial of Φ is readily calculated from the determinant form $\det(\Phi - \lambda I)$. Writing D_p for this determinant, one has:

$$\begin{vmatrix} \xi_1 - \lambda & \xi_2 & \xi_3 & \cdots & & & & \xi_p \\ 1 & -\lambda & 0 & \cdots & & & & 0 \\ 0 & 1 & -\lambda & \cdots & & & & 0 \\ 0 & 0 & 1 & \cdots & & & & 0 \\ \cdot & \cdot & \cdot & \cdots & & & & 0 \\ \cdot & \cdot & \cdot & \cdots & & & & 0 \\ & & & 0 & 1 & -\lambda & & 0 \\ & & & \cdot & 0 & 1 & & -\lambda \end{vmatrix}$$

By expanding the determinant in minors down the last column one finds:

$$D_p = (-)^{p-1}\xi_p \begin{vmatrix} 1 & -\lambda & 0 & 0 & \cdots & \\ 0 & 1 & -\lambda & 0 & \cdots & \\ \cdot & \cdot & \cdot & \cdot & \cdots & \\ \cdot & \cdot & \cdot & \cdot & \cdots & \\ & & & & 1 & -\lambda \\ & & & & 0 & 1 \end{vmatrix} - \lambda D_{p-1}$$

D_{p-1} being the same as D_p except for the deletion of the last column and row. The other determinant in the last equation is readily evaluated; multiplication of the last row by λ and addition of the result to the second-last row reduces the latter to a row whose only non-vanishing element is 1 in the diagonal position; repetition of the procedure upwards through the determinant finally leaves a unit determinant with 1 in every diagonal position and zero everywhere else. Hence the recursive relationship:

$$D_p = (-)^{p-1}\xi_p - \lambda D_{p-1}$$

follows, and one can write:

$$-D_2 = \xi_2 + \xi_1\lambda - \lambda^2$$

$$D_3 = \xi_3 + \xi_2\lambda + \xi_1\lambda^2 - \lambda^3$$

and the characteristic polynomial is found to be:

$$\xi_p + \xi_{p-1}\lambda + \xi_{p-2}\lambda^2 + \ldots + \xi_1\lambda^{p-1} - \lambda^p$$

Because of these restrictions, and the equivalence of periodicity and predictability by linear means, the methods discussed here are of limited value. A well-developed body of theory and numerical methodology exists for the extraction of periodic components from data, especially for data sampled at uniform intervals. It is, however, of some interest that the problem of prediction can be framed as an inversion problem, even though at first sight there is little apparent connection between these subjects.

Appendix

1. DETERMINANTS

Given a square array containing N rows and N columns of numbers, the associated determinant is written as:

$$\begin{vmatrix} a_{11} & a_{12} & a_{13} \cdots a_{1N} \\ a_{21} & a_{22} & a_{23} \cdots a_{2N} \\ \cdot & \cdot & \cdot \quad \cdot \\ \cdot & \cdot & \cdot \quad \cdot \\ a_{N1} & a_{N2} & a_{N3} \cdots a_{NN} \end{vmatrix}$$

or more concisely as det(**A**) or even $|\mathbf{A}|$, if **A** is the matrix symbol for the square array. The determinant is a single number (i.e. a scalar) formed from the $N \times N$ array as follows:

All combinations of N of the elements of the array are taken which contain one and only one element from each row and from each column. For a 3×3 array, for example, which contains nine elements (which would give eighty-four different possible combinations taken three at a time), only the following products enter into the determinant:

$a_{11}\,a_{22}\,a_{33}$, $\quad a_{11}\,a_{23}\,a_{32}$, $\quad a_{12}\,a_{21}\,a_{33}$, $\quad a_{12}\,a_{23}\,a_{31}$, $\quad a_{13}\,a_{21}\,a_{32}$

and $a_{12}\,a_{22}\,a_{31}$

— all other combinations contain more than one element from some row(s) or column(s). Since no row can contribute two elements to any given term, the determinant can be expressed in terms of combinations of elements which are ordered according to the row in which the corresponding element originated (i.e. according to the first subscript). Thus the determinant of an arbitrary square matrix **A** can be written:

$$\det(\mathbf{A}) = \sum(-)^s a_{1k_1} a_{2k_2} a_{3k_3} \cdots a \qquad (k_1 \neq k_2 \neq k_3 \neq k_4 \neq \ldots \neq k_N)$$

the sum being taken over all combinations of second subscript in which no such subscript is repeated. The sign $(-)^s$ is given by s, the number of direction reversals (i.e. $k_m < k_{m+1}$) in the sequence $k_1, k_2, \ldots, k$. Thus for $N = 5$ the combination $a_{11}\; a_{22}\; a_{34}\; a_{45}\; a_{53}$ contains a single reversal since the subscript sequence is 1, 2, 4, 5, 3 and so is assigned a negative sign, the combination $a_{12}\; a_{23}\; a_{31}\; a_{45}\; a_{54}$ is given a positive sign since there are two reversals (2, 3, 1, 5, 4).

Minors. Since the first row (for example) contributes only once to each combination entering into the determinant, the determinant can be expanded as a linear combination of the elements of the first row — or of the elements of any other row or column. Thus:

$$\det(\mathbf{A}) = a_{11}M_{11} + a_{12}M_{12} + a_{13}M_{13} + \ldots + a_{1N}M_{1N}$$

It can be shown that the coefficients M_{11}, M_{12}, etc., are given by $M_{mn} = (-)^{m+n-}$ multiplied by the "minor" determinant which is obtained by striking out the mth row and nth column of the original determinant. For example:

$$\begin{vmatrix} a_{11} & a_{12} & a_{13} \\ a_{21} & a_{22} & a_{23} \\ a_{31} & a_{32} & a_{33} \end{vmatrix} = a_{11} \begin{vmatrix} a_{22} & a_{23} \\ a_{32} & a_{33} \end{vmatrix} - a_{12} \begin{vmatrix} a_{11} & a_{13} \\ a_{31} & a_{33} \end{vmatrix} + a_{13} \begin{vmatrix} a_{21} & a_{22} \\ a_{31} & a_{32} \end{vmatrix}$$

as can be readily verified.

Properties of determinants. From the fundamental definition of a determinant certain properties necessarily follow:

(1) Exchanging a pair of rows, or a pair of columns changes the sign of the determinant, leaving it otherwise unchanged.

(2) Addition or subtraction (element by element) of a row to or from another row, or of a column to or from another column, does not change the value or the sign of the determinant.

(3) Any arbitrary linear combination of other rows can be added or subtracted from any given row without altering the determinant. The same holds true of column operations.

(4) If two rows or two columns of a determinant are equal (or equal apart from a reversal of sign) the determinant is zero.

2. MATRIX PROPERTIES INVOLVING DETERMINANTS

The eigenvalues $\lambda_1, \lambda_2, \ldots, \lambda_N$ of a matrix $\mathbf{A}$ were seen to be given by the equation:

$$\det(\mathbf{A} - \lambda\mathbf{I}) = 0$$

which can be expanded into a polynomial in λ:

$$\beta_0 + \beta_1\lambda + \ldots + \beta_N\lambda^N = 0$$

Setting $\lambda = 0$ gives:

$$\beta_0 = \det(\mathbf{A})$$

The eigenvalues $\lambda_1, \lambda_2, ..., \lambda_N$ are the roots of the characteristic polynomial, and one can write:

$$\beta_0 + \beta_1\lambda + \beta_2\lambda^2 + ... + \beta_N\lambda^N = (\lambda_1 - \lambda)(\lambda_2 - \lambda) ... (\lambda_N - \lambda)$$

Hence:

$$\beta_0 = \det(\mathbf{A}) = \lambda_1 \lambda_2 ... \lambda_N$$

In other words, the determinant of a matrix equals the product of its eigenvalues. A vanishing determinant implies a vanishing eigenvalue and vice versa.

The term involving β_1 can be isolated by differentiating the determinant. To differentiate a determinant with respect to a variable x, one differentiates the first row (or column) term by term, holding all other rows (or columns) constant; this contributes one term to the derivative:

$$\begin{vmatrix} \frac{da_{11}}{dx} & \frac{da_{12}}{dx} & \cdots & \frac{da_{1N}}{dx} \\ a_{21} & a_{22} & \cdots & a_{2N} \\ \cdot & \cdot & & \\ \cdot & \cdot & & \\ a_{N1} & a_{N2} & \cdots & a_{NN} \end{vmatrix}$$

The whole derivative is the sum of N such terms, obtained by proceeding through all rows (or columns) in a similar way.

Differentiation of $\det(\mathbf{A} - \lambda\mathbf{I})$ with respect to λ gives:

$$\beta_1 = \begin{vmatrix} -1 & 0 & 0 & \cdots & 0 \\ a_{21} & a_{22} & a_{23} & \cdots & a_{2N} \\ \cdot & \cdot & \cdot & & \cdot \\ \cdot & \cdot & \cdot & & \cdot \\ a_{N1} & a_{N2} & a_{N3} & \cdots & a_{NN} \end{vmatrix} + \begin{vmatrix} a_{11} & a_{12} & a_{13} & \cdots & a_{1N} \\ 0 & -1 & 0 & \cdots & 0 \\ \cdot & \cdot & \cdot & & \cdot \\ \cdot & \cdot & \cdot & & \cdot \\ a_{N1} & a_{N2} & a_{N3} & \cdots & a_{NN} \end{vmatrix} + \cdots$$

$$= -M_{11} - M_{22} - M_{33} - ... - M_{NN}$$

The term in λ^N is $\det(\mathbf{A} - \lambda\mathbf{I})(-)^N \lambda^N$. The term in λ^{N-1} is evidently $(-)^{N-1}(\lambda_1 + \lambda_2 + ..., + \lambda_N)\lambda^{N-1}$. To obtain terms in λ^{N-1} it is clear that all but one of the diagonal elements of the determinant $\det(\mathbf{A} - \lambda\mathbf{I})$ *must* be taken; if all diagonal elements but the rth are taken, all but the rth column and rth rows have then been used up and the only permissible element with which they can be combined is in fact the diagonal element. The term in λ^{N-1} is therefore contributed to only by the product of diagonal elements:

$(a_{11} - \lambda)(a_{22} - \lambda) \ldots (a_{NN} - \lambda)$

and hence is $(-)^{N-1} \lambda^{N-1} (a_{11} + a_{22} + \ldots + a_{NN})$. Hence:

The sum of the diagonal elements of **A** equals the sum of the eigenvalues. The sum of the diagonal is known as the *trace* of the matrix.

3. SOLUTION BY DETERMINANTS OF A LINEAR SYSTEM OF EQUATIONS

It can be shown that a linear system of N equations in N unknowns x_1, $x_2, \ldots, x_N$, of the form:

$$\left.\begin{array}{l} a_{11}x_1 + a_{12}x_2 + \ldots + a_{1N}x_N = y_1 \\ a_{21}x_1 + a_{22}x_2 + \ldots + a_{2N}x_N = y_2 \\ \cdot \\ \cdot \\ a_1x_1 + a_2x_2 + \ldots + a_{NN}x_N = y_N \end{array}\right\}$$

is solved by:

$$x_1/D_1 = x_2/D_2 = \ldots = x/D = 1/\Delta$$

where the determinants formed from the $N \times (N + 1)$ array by striking out the kth column of a's; D_2, for example, is:

$$\begin{vmatrix} a_{11} & a_{13} & a_{14} \ldots a_{1N} & y_1 \\ a_{21} & a_{23} & a_{24} \ldots a_{2N} & y_2 \\ \cdot & \cdot & \cdot & \\ \cdot & \cdot & \cdot & \\ a_{N1} & a_{N3} & a_{N4} \ldots a_{NN} & y_N \end{vmatrix}$$

Δ is the determinant of the array of a's, i.e.

$$\Delta = \det(\mathbf{A}) = \begin{vmatrix} a_{11} & a_{12} \ldots a_{1N} \\ a_{21} & a_{22} \ldots a_{2N} \\ \cdot & \cdot \\ \cdot & \cdot \\ a_{N1} & a_{2N} \ldots a_{NN} \end{vmatrix}$$

Evidently for a homogeneous system in which the y's are all zero, a non-zero solution for the x_k is possible *only* when $\Delta = 0$.

Suggestions for further reading

Inversion mathematics

Baker, C.T., Fox, L., Mayers D.F. and Wright, K., 1964. *Compt. J.*, 7: 141—148.

Berezin, I.S. and Zhidkov, N.P., 1965. *Computing Methods.* Pergamon, London, 620 pp.

Fox, L. and Goodwin, E.T., 1953. *Philos. Trans. R. Soc. Lond., Ser. A*, 241: 501—534.

Hanson, R.J., 1971. *Soc. Ind. Appl. Math., J. Numer Anal.*, 8: 616—622.

Lawson, C.L. and Hanson, R.J., 1974. *Solving Least Squares Problems.* Prentice Hall, Englewood Cliffs, N.J., 340 pp.

Strand, O.N. and Westwater, E.R., 1968. *Soc. Ind. Appl. Math., J. Numer. Anal.*, 5: 287—295.

Tikhonov, A.N., 1963. *Sov. Math. Dokl.*, 5: 835—838.

Varah, J.M., 1973. *Soc. Ind. Appl. Math., J. Numer. Anal.* 10: 257—267.

Integral equations

Mikhlin, S.G., 1965. *Multidimensional Singular Integrals and Integral Equations.* Pergamon, Oxford, 259 pp.

Robinson, S.M. and Stroud, A.H., 1961. *Math. Comput.*, 15: 286—288;

Smithies, F., 1958. *Integral Equations.* Cambridge University Press, Cambridge, 172 pp.

Sneddon, I.N., 1955. Functional analysis. In: *Handbuch der Physik, Mathematical Methods II.* Springer, Berlin, pp. 198—348.

Sneddon, I.N., 1972. *The use of Integral Transforms.* McGraw-Hill, New York, N.Y., 539 pp.

Fourier integrals and transforms

Bakeen, W.H. and Eisenberg, A., 1970. *Bull. Seismol. Soc. Am.*, 60: 1291—1296.

Cooley, J.W. and Tukey, J.W., 1965. *Math. Comput.*, 19: 297—301.

Erdelyi, A., 1954. *Tables of Integral Transforms.* McGraw-Hill, New York, N.Y., 451 pp.

Gold, B. and Rader, C.M., 1969. *Digital Processing of Signals.* McGraw-Hill, New York, N.Y., 269 pp.

Lighthill, M.J., 1958. *Introduction to Fourier Analysis and Generalized Functions.* Cambridge University Press, Cambridge, 79 pp.

Natanson, I.P., 1964. *Constructive Function Theory*, Ungar, New York, N.Y., Vol. 1, 514 pp.

Sanger, A. 1964. *J. Math. Anal. Appl.*, 8: 1—2.

Inversion of Laplace transforms

Atkinson, M.P. and Lang, S.R., 1972. *Comput. J.* 15: 138—139.

Bellman, R.E., Kalaba, R. and Lockett, J.A., 1966. *Numerical Inversion of the Laplace Transform. Applications to Biology, Economics, Engineering and Physics.* American Elsevier, New York, N.Y., 249 pp.

Dubner, H. and Abate, J., 1968. *J. Assoc. Comput. Mach.*, 15: 115—123.

Krylov, V.I. and Skoblya, N.S., 1969. *Handbook of Numerical Inversions of Laplace Transforms.* Israel Program for Scientific Translation, Jerusalem, 293 pp.

Piessens, R. and Branders, M., *Proc. Inst. Electr. Eng.*, 118: 1517—1522.

Salzer, H.E., 1958. *J. Math. Phys.*, 37: 89—108.
Shirtliffe, C.J. and Stephenson, D.G., 1961. *J. Math. Phys.*, 40: 135—141.
Spinelli, R.A., 1966. *Soc. Ind. Appl. Math., J. Numer. Anal.*, 3: 636—649.
Twomey, S., 1963. *J. Franklin Inst.*, 275: 121—138.
Weeks, W.T., 1966. *J. Assoc. Comput. Mach.*, 13: 419—429.
Zakian, V., 1970. *Electron. Lett.*, 6: 677—679.

Numerical quadrature

Abramovitz, M., 1954. *Soc. Ind. Appl. Math., J. Appl. Math.*, 2: 20—35.
Chakravarti, P.C., 1970. *Integrals and Sums: Some New Formulae for their Numerical Evaluation.* Athlone, London, 89 pp.
Davis, P.J. and Rabinovitz, P., 1967. *Methods of Numerical Integration.* Academic Press, New York, N.Y., 459 pp.
Dixon, V.A., 1974. Numerical quadrature: a survey of available algorithms. In: D.J. Evans (Editor), *Software For Numerical Mathematics.* Academic Press, New York, N.Y., pp. 105—137.
Ghizetti, A. and Ossicini, A., 1970. *Quadrature Formulae.* Academic Press, New York, N.Y., 192 pp.
Hamming, R.W. and Epstein, M.P., 1972. *Soc. Ind. Appl. Math., J. Numer. Anal.*, 9: 464—475.
Krylov, V.I., 1962. *Approximate Calculation of Integrals* (translation, A.H. Stroud). Macmillan, New York, N.Y., 357 pp.
Nikolskii, S.M., 1966. *Quadrature Formulae.* H.M. Stationery Office, London, 191 pp.
Ralston, A., 1965. *A First Course in Numerical Analysis.* McGraw-Hill, New York, N.Y., 578 pp.
Squire, W., 1970. *Integration for Engineers and Scientists.* American Elsevier, New York, N.Y., 302 pp.
Stroud, A.H., 1971. *Approximate Calculation of Multiple Integrals.* Prentice Hall, Englewood Cliffs, N.J. 431 pp.
Stroud, A.H. and Secrest, D.H., 1966. *Gaussian Quadrature Formulae.* Prentice Hall, Englewood Cliffs, N.J., 374 pp.

Remote sensing in solid-earth geophysics

Al-Chalabi, M., 1971. *Geophysics,* 36: 835—855.
Al-Chalabi, M., 1972. *Geophys. Prospect.*, 20: 1—16.
Banks, R.J., 1972. *J. Geomagn. Geoelectr.*, 24: 337—351.
Banks, R.J., 1973. *Phys. Earth Planet. Inter.*, 7: 339—348.
Bullen, K.E., 1960. *Geophys. J. R. Astron. Soc.*, 3: 258—269.
Cordell, L. and Henderson, R.G., 1968. *Geophys,* 33: 596—601.
Grant, F.S., 1972. *Geophysics,* 37: 647—661.
Ku, C.C., Telford, W.M. and Lim, S.H. 1971. *Geophysics,* 36: 1174—1203.
Kuratnam, K., 1972. *Geophys. Prospect.*, 20: 439—447.
Paul, M.K., 1967. *Geophysics,* 32: 708—719.
Parker, R.L. 1972. *Geophys. J. R. Astron. Soc.*, 29: 123—138.
Price, A.T., 1973. *Phys. Earth Planet. Inter.*, 7: 227—233.
Qureshi, I.R. and Idries, F.M., 1972. *Geophys. Prospect.*, 20: 106—108.
Roy, A., 1966. *Geophys,* 31: 167—184.
Sax, R.L., 1966. *Geophysics,* 31: 570—575.
Steinhart, J.S. and Meyer, R.P. 1961. *Explosion Studies of Continental Structure.* Carnegie Institute of Washington Publ. 622, 409 pp.
Wiggins, R.A., 1972. *Rev. Geophys. Space Phys.*, 10: 251—285.

Remote sensing of the atmosphere

Derr, V.E. (Editor), 1972. *Remote Sensing of the Troposphere.* Government Printing Office, Washington, D.C., 613 pp.
Gorchakova, I.A., Malkevitch, M.C. and Turchin, V.E., 1970. *Izv. Atmos. Ocean Phys.*, 6: 565—576.
Hanel, R.A. and Conrath, B., 1969. *Science*, 165: 1258—1260.
Houghton, J.T., 1961. *Q J. R. Meteoreol. Soc.*, 87: 102—104.
Holz, R.K. (Editor), 1973. *The Surveillance Science: Remote Sensing of the Environment.* Houghton-Mifflin, Boston, Mass., 390 pp.
Kaplan, L.D., 1959. *J. Opt. Soc. Am.*, 49: 1004—1007.
King, J.I.F., 1963. *J. Atmos. Sci.*, 20: 245—250.
Kondratiev, K.Y., Volynov, B.V., Galtsev, A.P. Smotky, O.I. and Khrunov, E.V., 1971. *Appl. Opt.*, 10: 2521—2533.
Lusignan, B., Medrell, G., Morrison, A., Polamaza; J. and Ungar, S.G., 1969. *Proc. Inst. Electr. Electron. Eng.*, 57: 458—467.
National Academy of Sciences U.S., 1969. *Atmospheric Exploration by Remote Probes.* New York, N.Y., 698 pp.
Smith, W.L. and Howell, H.B., 1971. *J. Appl. Meterol.*, 10: 1026—1034.
Singer, S.F. and Wentworth, R.C. 1957. *J. Geophys. Res.*, 62: 299—308.
Staelin, D.H., 1969. *Proc. Inst. Electr. Electron. Eng.*, 57: 427—439.
Twomey, S. and Howell, H.B., 1963. *Mon. Weather Rev.*, 91: 659—664.
Wark, D.Q., 1970. *Appl. Opt.*, 9: 1961—1970.
Westwater, E.R., 1972. *Mon. Weather Rev.*, 100: 15—28
Yates, H.W., 1970. *Appl. Opt.*, 9: 1971—1975.

Remote sensing — other applications

Brinkmann, R.T., 1976. *Icarus*, 27: 69—79.
Fjeldbo, G. and Eshleman, V.R., 1965. *J. Geophys. Res.*, 70: 3217—3225.
Krakow, B., 1966. *Appl. Opt.*, 5: 201—209.
Phinney, R.A. and Anderson, D.L., 1968. *J. Geophys. Res.*, 73: 1819—1827.
Twomey, S., 1976. *J. Atmos. Sci.*, 33: 1073—1079.
Yamamoto, G. and Tanaka, M., 1969. *Appl. Opt.*, 8: 447—453.

Name index

Subject index